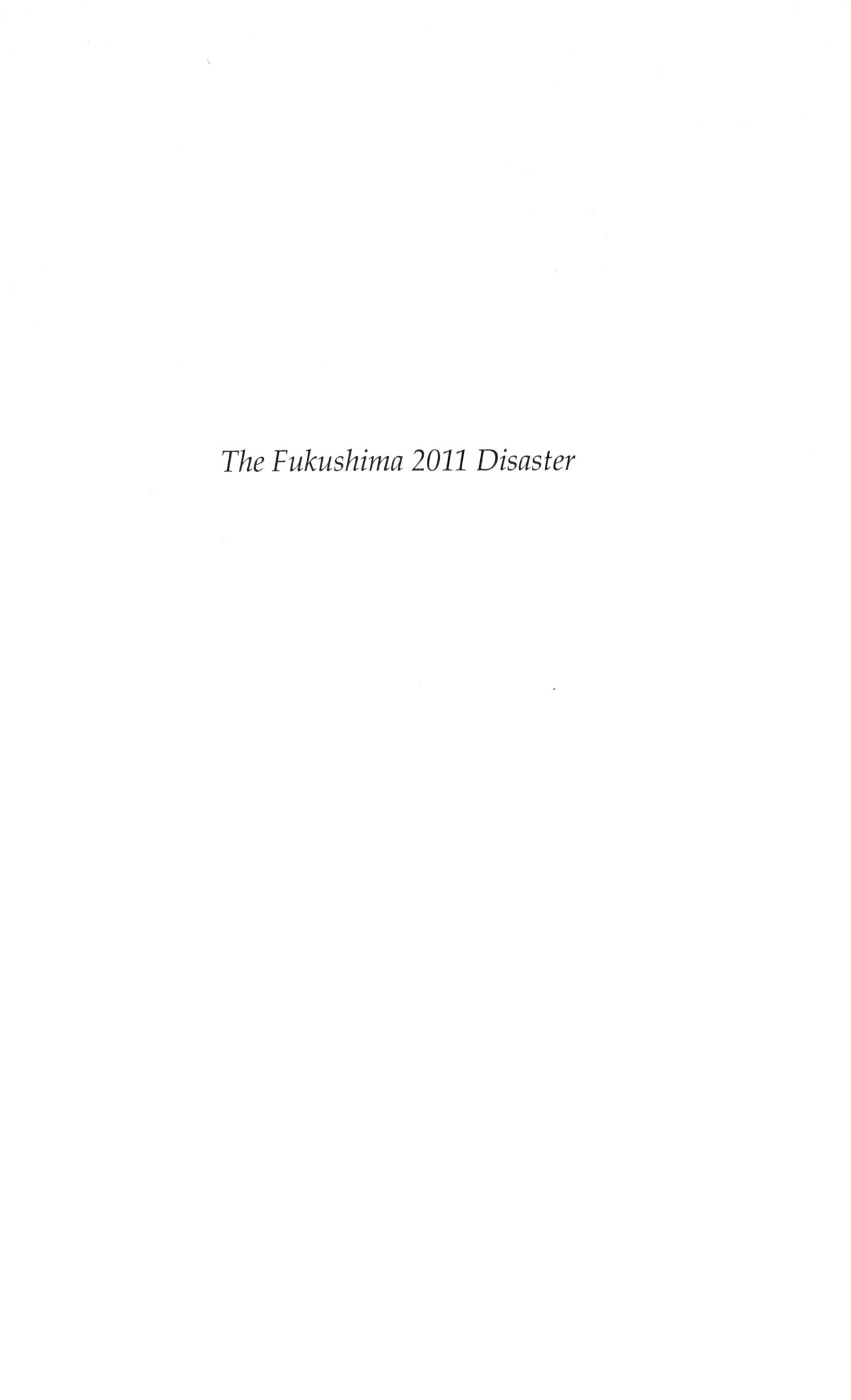

*The Fukushima 2011 Disaster*

# The Fukushima 2011 Disaster

Ronald Eisler
Senior Scientist *(retired)*
U.S. Geological Survey

CRC Press is an imprint of the
Taylor & Francis Group, an **informa** business
A SCIENCE PUBLISHERS BOOK

CRC Press
Taylor & Francis Group
6000 Broken Sound Parkway NW, Suite 300
Boca Raton, FL 33487-2742



Printed in the United States of America on acid-free paper

Version Date: 20120327

International Standard Book Number: 978-1-4665-7782-4 (Hardback)



**Visit the Taylor & Francis Web site at**
**http://www.taylorandfrancis.com**

**CRC Press Web site at**
**http://www.crcpress.com**

**Science Publishers Web site at**
**http://www.scipub.net**

## *Dedication*

*This book is dedicated to my family*
*Jeannette, Reneé, David, Charles, Julie and Eb.*

# Preface

In 1945, about 153,000 Japanese were killed and more than 100,000 injured when nuclear bombs were detonated over Hiroshima and Nagasaki. By January 2011, Japan had embraced nuclear technology with 54 operating reactors and ranked among the wealthiest developed nations. The earthquake of March 2011, followed by a massive tsunami resulted in the deaths of at least 20,000 people. The tsunami inundated the nuclear facilities at Fukushima, causing explosions, the release of huge, and still ongoing, amounts of radioactive materials which resulted in the forced evacuation of more than 300,000 residents. At present, there are more than 400 nuclear reactors in operation globally with a major nuclear event predicted every 8 years. This book attempts to critically analyze the actions taken by Japan and other nations over the next 15 months in response to the Fukushima disaster so as to prevent future occurrences.

# Acknowledgements

Research materials were kindly provided by librarians of the U.S. Department of the Interior in Laurel, Maryland, the National Agricultural Library in Beltsville, Maryland, and the Nuclear Regulatory Commission in Rockville, Maryland. Illustrations were provided by Julie Dietrich-Eisler and Katie O'Meara. I thank Vijay Primlani of Science Publishers for expediting publication.

# Contents

# List of Figures

# List of Tables

# Introduction

## 1.1 General

The Fukushima 1 nuclear power plant, also known as Fukushima Dai-ichi (Dai-ichi means "number one") is a disabled nuclear power plant located on a 3.5 square km (860 acres) site in the towns of Okuma and Futaba in the Futaba district of Fukushima Prefecture, Japan (Wikipedia 2011b; Figure 1.1). First commissioned in 1971, the plant consists of six boiling water reactors. These light-water reactors drove electric generators with a combined power of 4.7 GWe (4.7 billion electrical watts), making the Fukushima facility the 15th largest nuclear power complex in the world and the first to be constructed and run entirely by the Tokyo Electric Power Company (TEPCO). The plant suffered major damage from an earthquake and subsequent tsunami on March 11, 2011 and is not expected to reopen. On April 20th, Japanese authorities declared a 20-km evacuation zone around the facility which may be entered only under government supervision (Wikipedia 2011b).

Fukushima, which means "fortunate island", is the home of the first Japanese nuclear facility and was erected on the former site of a World War II imperial air base; later, another facility was built nearby (Osnos 2011). On March 11, 2011, a 9.0 magnitude earthquake struck the east coast of Japan (Becker 2011; Christodouleas et al., 2011; Dauer et al., 2011; Hall 2011; Kuczera 2011; Matanle 2011; Miller et al., 2011; Mimura et al., 2011; Osnos 2011; Wikipedia 2011; Anzai et al., 2012; Ohnishi 2012). The total number of people who died in the earthquake and the 13 to 15-meter high tsunami that it generated was at least 14,000

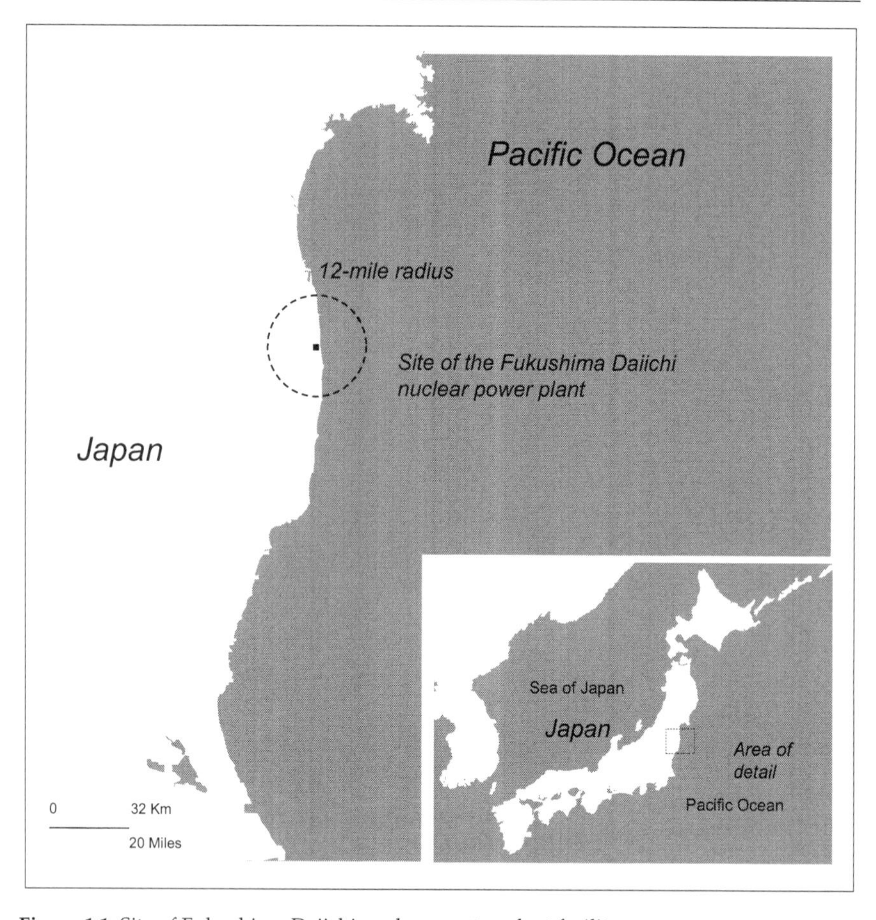


**Figure 1.1.** Site of Fukushima Daiichi nuclear reactor plant facility.

and may be as high as 25,000. Structural damage to the six reactors at Fukushima Dai-ichi is documented with significant and continuing radiation losses to the environment. More than 200,000 inhabitants from the vicinity of the site have been evacuated, with some estimates ranging as high as 320,00—some never to return owing to continuing radiation hazards. The status of the facility continues to change and permanent control of its radioactive inventory has not been achieved. At present, the situation at the Fukushima nuclear facility remains fluid and the long-term environmental and health impact will probably take years—if not decades—to fully evaluate. Japanese authorities have responded to the event through organized evacuation from the vicinity of the site; monitoring of food and water; establishing radiation limits on such foodstuffs; distribution of stable potassium iodide;

and systematic scanning of evacuees. As night fell on March 11th, it became clear that TEPCO—which produces about one-third of the country's electricity—had been unprepared for such a disaster (Becker 2011; Christodouleas et al., 2011; Dauer et al., 2011; Hall 2011; Kuczera 2011; Matanle 2011; Miller et al., 2011; Mimura et al., 2011; Osnos 2011; Wikipedia 2011; Anzai et al., 2012).

The reactors for Units 1, 2 and 6 were supplied by General Electric, those for Units 3 and 5 by Toshiba and Unit 4 by Hitachi (Wikipedia 2011b). Since September 2010, Unit 3 was fueled by a small fraction of mixed-oxide fuel rather than the low enriched uranium used in the other reactors. Unit 1 is a 460 MW boiling water reactor constructed in July 1967. It commenced commercial electrical production on March 16, 1971 and was initially scheduled for shutdown in early 2011. In February 2011, Japanese regulators granted an extension of ten years for continued operation of the reactor. All units were inspected after the 1978 Miyagi earthquake, although no damage to reactor critical parts was discovered. The design basis for tsunamis was 5.7 m and a seawall of that height was erected around the facility (Wikipedia 2011b).

Units 1 though 6 came online from 1970 through 1979 (Wikipedia 2011b). From 2002 through 2005, the reactors were shut down for safety checks after TEPCO admitted that it had supplied fake inspection and repair reports to the Japanese Nuclear and Industrial Safety Agency. The report revealed that TEPCO failed to inspect more than 30 technical components of the six reactors, including power boards for the reactors' temperature control valves, water pump motors and emergency diesel generators. In 2008, the International Atomic Energy Agency (IAEA 2011) warned Japan that the Fukushima reactors were built using outdated safety guidelines and could present a serious problem during a large earthquake. The warning led to the building of an emergency response center in 2010 and was in use in the 2011 event. Planned construction of two additional reactors, Units 7 and 8, were scheduled to begin in April 2012 and 2013 and to go online in October 2016 and 2017, respectively; however, the project was formally cancelled by TEPCO in April 2011 (Wikipedia 2011b).

In the days that followed, reactors 1, 2 and 3 experienced full meltdown (Wikipedia 2011c; Anzai et al., 2012). The upper 77% of the core of unit 1 melted and slumped in the lower quarter of the core on March 12th. Hydrogen explosions destroyed the upper cladding of the buildings housing reactors 1, 3 and 4, with additional damage to reactor 2. Unit 1—and the other two melted down reactors—continued to leak cooling water for at least three months after the initial event. Fuel rods stored in pools in each reactor building began to overheat as water levels in the pools dropped. Fears of radioactivity releases led to a 20-km (12-mile) radius evacuation around the plant. Measurements taken by Japanese scientists in areas of northern Japan 30 to 50 km from the plant showed unacceptably high levels of radiocesium; foods grown in this area were banned from sale. Tokyo officials recommended that tap water should not be used to prepare food for infants, at least temporarily. The Japanese government and TEPCO have been criticized in the foreign press for poor communication with the public and improvised cleanup efforts. Foreign experts have said that a workforce in the hundreds or even thousands would take 20 years or more to clean up the area (Harlan 2011c); however, on March 20th, the Chief Cabinet Secretary announced that the plant would be decommissioned once the crisis was over (Wikipedia 2011c).

Since March 11, 2011, when an earthquake and tsunami struck Japan, affecting nuclear reactors at the Fukushima site, the U.S. Nuclear Regulatory Commission fully activated its 24-hour Emergency Operations Center to monitor and analyze events at the Japanese nuclear plants and to support the Japanese government and the U.S. ambassador in ensuring protection of health and safety of American nationals (U.S. NRC 2011).

All governments regulate risky industrial systems such as nuclear power plants in hopes of making them less risky including formal and informal warning systems to avoid catastrophe (Perrow 2011). All segments of society respond when disaster occurs, although recent history is rife with major disasters accompanied by failed regulation, ignored warnings, inept disaster response and human error. Despite all attempts to forestall them, accidents will inevitably occur in the tightly coupled systems of modern society, resulting in the kind of unpredictable, cascading disaster seen at the Fukushima nuclear power station. Government and business can always do more to prevent serious accidents through regulation, design and training; however, some complex systems with catastrophic potential are just too dangerous to exist, because they cannot be made safe regardless of human effort (Perrow 2011).

On November 20th, eight months later, the area surrounding the Fukushima nuclear plant remains a wasteland after the events of March 11th (Harlan 2011f). On December 16th, the Japanese government stated that the Fukushima reactors had reached a stable state known as "cold shutdown", a benchmark for progress in the 40-year effort to decommission the reactors (Harlan and Mie 2011). On December 21st Japan unveiled its plan to to dismantle the Fukushima nuclear plant and return the surrounding region to normal, a 40-year process that represents one of the most ambitious nuclear cleanups in history (Dvorak and Obe 2011). The first stage (2013-2014) is to be devoted to clearing debris from around the spent-fuel pools. The second stage (2015–2022) is, in order: start removing fuel from spent fuel rods; finish repair of reactor/turbine buildings; finish draining reactor/turbine buildings. The final stage (2023–2053) involves removing fuel from damaged reactors and the dismantling of reactor buildings. The program aims to bring the area around the nuclear plant back to pre-March 3rd levels as well as plans to decontaminate surrounding areas to radiation levels well below levels known to be dangerous so that people will feel safe moving back into it (Dvorak and Obe 2011). The Fukushima cleanup is expected to be a lengthy and costly undertaking (Stone 2011).

The current work summarizes and critically analyzes the natural events and human shortcomings responsible for the failure of the Fukushima reactors during the first 14 months following the accident and governmental and civilian responses to the emergency.

## LITERATURE CITED

Anzai, K., N. Ban, T. Ozawa and S. Tokonami. 2012. Fukushima Daiichi nuclear power plan accident: Facts, environmental contamination, possible biological effects and countermeasures, *Journal of Clinical Biochemistry and Nutrition*. **50(1)**:2–8.

Becker, S.M. 2011. Learning from the 2011 Great East Japan disaster: insights from a special radiological emergency assistance mission, *Biosecurity and Bioterrorism: Biodefense Strategy, Practice and Science*. **9(4)**:394–404.

Christodouleas, J.P., R.D. Forrest, C.G. Ainsley, Z. Tochner, S.M. Hahn and E. Glatstein. 2011. Short-term and long-term health risks of nuclear-power plant accidents, *New England Journal of Medicine*. **364**:2334–2341.

Dauer, L., P. Zanzonico, R.M. Tuttle, D.M. Quinn and H.W. Strauss. 2011. The Japanese tsunami and resulting nuclear emergency at the Fukushima Daiichi power facility: Technical, Radiologic and Response Perspectives, *Journal of Nuclear Medicine*. **52**:1423–1432.

Dvorak, P. and M. Obe. 2011a. Panel finds serious errors in Japan, *Wall Street Journal* (newspaper), December 27th, A7.

Hall, H.L. 2011. Fukushima Daiichi: Implications for carbon-free energy, nuclear nonproliferation and community resilience, Integrated Environmental Assessment and Management, 7, pp. 406–408, doi:10.1002/ieam.225.

Harlan, C. 2011c. Small hot spots in Tokyo heighten worry about radiation spread, Washington Post (newspaper), October 14th, A 7.

Harlan, C. 2011f. Into Japan's dead zone, Washington Post (newspaper), November 20th, A 26.

Harlan, C. and A. Mie. 2011. Japan: Damaged nuclear plant has been stabilized, Washington Post (newspaper), December 17th, A 5.

IAEA (International Atomic Energy Commission). 2011. Fukushima nuclear accident update log, iaea.org/...tsunamiupdate01.html

Kuczera, B. 2011. The severe earthquake in Tohoku in Japan and the impact on the Fukushima-Daiichi nuclear power station, *International Journal of Nuclear Power*. **56:**234–241.

Matanle, P. 2011. The Great East Japan earthquake, tsunami and nuclear meltdown: Towards the (re)construction of a safe, sustainable and compassionate society in Japan's shrinking regions, *Local Environment*. **16(9):**823–847.

Miller, C., A. Cubbage, D. Dorman, J. Grobe, G. Holahan and N. Sanfillippo. 2011. Recommendations for enhancing reactor safety in the 21st century. The near-term task force review of insights from the Fukushima Dai-ichi accident, Available from the U.S. Nuclear Regulatory Commission as document ML 111861807, 83 p.

Mimura, N., K. Yasuhara, S. Kawagoe, H. Yokoki and S.O. Kazama. 2011. Damage from the Great East Japan earthquake and tsunami—a quick report, *Mitigation and Adaptation Strategies for Global Change*. **16 (7):**803–818.

Ohnishi, T. 2012. The disaster at Japan's Fukushima-Daiichi nuclear power plant after the March 11, 2011 earthquake and tsunami and the resulting spread of radioisotope contamination, *Radiation Research*. **177 (1):**1–14.

Osnos, E. 2011. Letter from Fukushima. The fallout. The New Yorker, October **17:**46–61.

Perrow, C. 2011. Fukushima and the inevitability of accidents. *Bulletin of the Atomic Scientists*. **67(6):**44–52.

Stone, R. 2011. Fukushima cleanup will be drawn out and costly, *Science*. **331(6024):**1507.

U.S. NRC (U.S. Nuclear Regulatory Commission). 2011. Japan nuclear accident—NRC actions, September 23rd, http://www. nrc.gov/japan/japan-info. html/journal.pone.002761.

Wikipedia. 2011, Timeline of the Fukushima Daiichi nuclear disaster, 21 p.

Wikipedia. 2011b. Fukushima Daiichi nuclear power plant, 7 p.

Wikipedia. 2011c. Fukushima Daiichi nuclear disaster, 54 p.

Chapter 2

# The Earthquake

## 2.1 General

Mid-continent events, far from plate boundaries such as the so called Ring of Fire along which California trembles, are poorly understood, but can be just as destructive (Maynard 2012). For example, the 1556 disaster in Shaanxi, China, had the highest death toll in history, an estimated 830,000. Earthquakes happen along meandering faults and are unpredictable. And they can be widely separated in time. The August 2011 magnitude 5.8 earthquake at Mineral, Virginia, which damaged the Washington Monument and other buildings in Washington, D.C., was the biggest in that seismic zone since 1875. The U.S. Geological Survey (USGS) continues to warn of the potential for a major destructive earthquake from Memphis, Tennessee, north to southern Illinois, as based on the history of past earthquakes. The USGS estimates the chance of having a magnitude 7 to 8 mid-continent earthquake during the next 50 years is 7 to 10 percent and the chance of having a magnitude 6 or larger earthquake in 50 years is 25 to 40 percent (Maynard 2012).

The 96 nuclear reactors in the central and eastern United States face previously unrecognized threats from big earthquakes, according to a report of the U.S. Nuclear Regulatory Commission (NRC) in late January 2012 (Smith 2012). The NRC plans to give operators of nuclear plants a period of four years to reevaluate seismic risks for all structures, systems and components according to a complex model jointly developed by the

NRC, the U.S. Department of Energy and the industry-funded Electric Power Research Institute. The model incorporates data from about 1,000 earthquakes that were previously not cataloged, bringing the total for the region to nearly 3,300 earthquakes since 1568. Retrofitting old nuclear plants to conform to the model and to withstand the single worst earthquake most likely to occur in 10,000 years, however, may be so costly that the reactors might be closed altogether (Smith 2012).

## 2.2 Japan

Pacific rim countries, especially Japan, are particularly vulnerable to tsunamis which—in the past—have caused widespread destruction of property and life (National Geographic 2005).

On March 11th, 2011, a powerful magnitude 9.0 quake (as measured on the Richter scale) occurred 373 km northeast of Tokyo and 130 km east of Sendai on the main island of Honshu—which is located in the Pacific Ocean—at a depth of 24.4 km (Diep 2011; Figure 2.1). This was the 5th most powerful recorded earthquake in the world since 1900 and the biggest ever measured in the history of Japan. More than 1,200 years have passed since Japan has experienced an earthquake approaching this magnitude. The Pacific Plate struck the 1,300 km long by 80 km wide Honshu with sufficient force to move the island 2.4 m and about 3 cm closer to the United States in less than a week. The duration of strong shaking reported from Japan during the March 11th quake ranged from three to five minutes Prior to the main quake, there were 4 foreshocks of magnitudes 6.0, 6.1. 6,1 and 7.2 respectively, with the worldwide annual average of earthquakes of magnitude greater than 6.0 being 150. There were at least 401 confirmed aftershocks (Diep 2011). At the time of the quake (first detected at 2:46 PM), about 6,000 workers were inside the plant, 246 km from Tokyo (Osnos 2011). Doors jammed as heavy objects fell, but loss of life that was earthquake-related was insignificant. On that same day Reactors 4, 5 and 6 of the six Fukushima plant's reactors were down for routine maintenance and the other 3 shut down automatically in response to the earthquake (Osnos 2011). Based on multi-spectral remote sensing data, 76% of all buildings in Miyagi Prefecture were affected by seismic activity 6 and above, with lesser values for Fukushima and Iwate (Guo et al., 2011)

The tremor caused the power plant to be cut off from the Japanese electricity grid; however, backup diesel generators provided power to the emergency condenser designed to cool the steam inside the pressure vessel if the reactor fails (Wikipedia 2011). On April 11th,

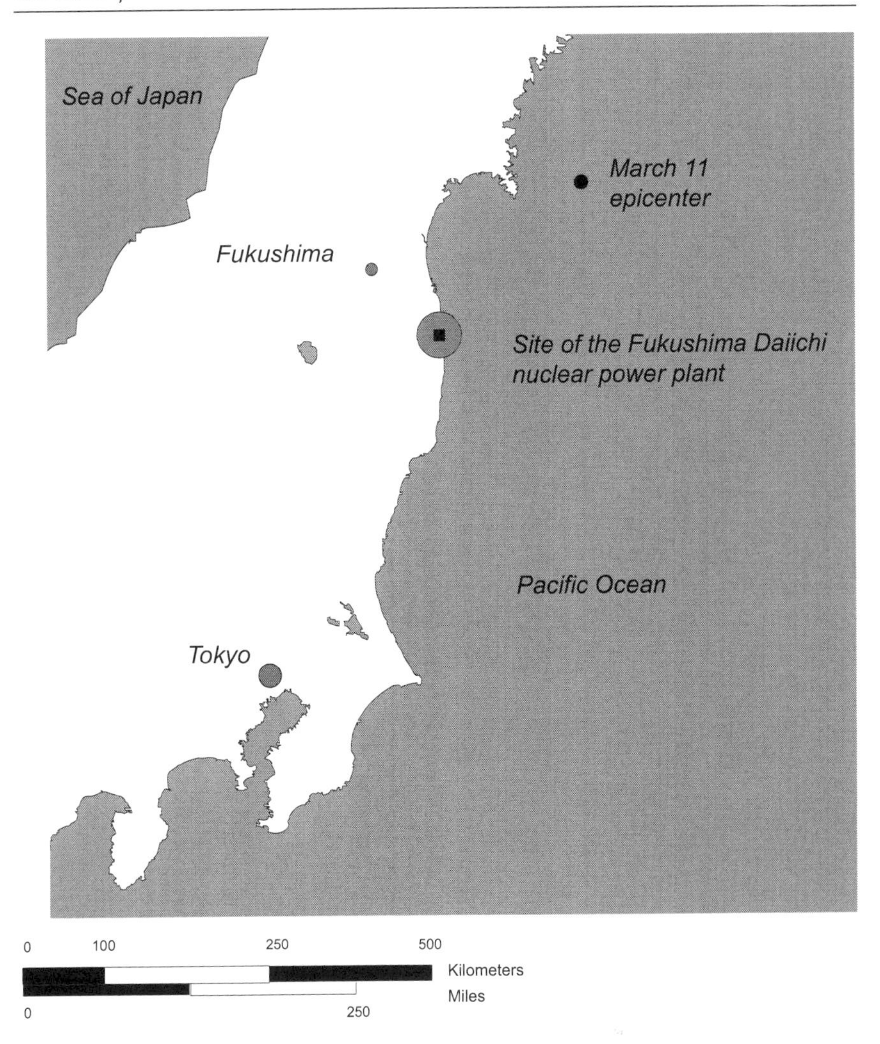


**Figure 2.1.** Earthquake epicenter and zones of no-entry and expanded evacuation surrounding Fukushima.

*(Color image of this figure appears in the color plate section at the end of the book.)*

coolant injection into reactors 1 and 3 was interrupted for 50 minutes after a strong aftershock (Wikipedia 2011). The April 11th aftershock, measuring, 7.0, caused numerous fault ruptures on land (Fisher 2011; Miyagi et al., 2011). The Japanese Landslide Society noted the failure of a water reservoir embankment dam in Sukagawa, Fukushima prefecture, landslides and surface seismic fault ruptures from the April 11th aftershock in Iwaki, Fukushima, a concentration of surface failures at

Matsushima Bay in Miyagi prefecture and small landslides on modified slopes in residential areas around Sendai city (Miyagi et al., 2011). The rupture process of the mega-thrust earthquake is complex and was investigated using low-frequency strong-motion records, epicentral distances, various fault models, acceleration waveforms and frequency of the seismic wave radiation (Suzuki et al., 2011).

A powerful and independent panel of specialists appointed by Japan's parliament is challenging the Government's account of the accident at Fukushima, including an inquiry into how much the March earthquake may have damaged the plant's reactor even before the tsunami struck (Tabuchi 2012). One panel member argued that the earthquake was likely to have damaged Fukushima reactors at the plant to the extent that melt-downs would have occurred without the tsunami, a point disputed by TEPCO (Tabuchi 2012). According to the Japanese government's forecast, there is a 70% chance of a magnitude-7 earthquake hitting Tokyo over the next 30 years (Obe 2012a). However, a new research investigation, using different methodology, has determined that there is a 70% chance of a magnitude-7 earthquake striking the Tokyo metropolitan area within only four years and a 98% chance over the next 30 years (Obe 2012a).

Earthquake prediction remains an inexact science. The Earthquake Research Institute of Tokyo University and other Japanese organizations have received billions dollars for earthquake prediction (Harlan 2012). Experts have spent decades looking for earthquake precursors by examining changes in the levels of water and radon gas, electrical current, animal behavior, historical data, fault lines, readings from attached instruments to the sea floor and pulverized rocks in the laboratory—all in a search for clues about how the earth behaves before it slips, shifts and shakes. But none of the major earthquakes to hit Japan since 1979 have occurred in areas that government seismologists describe as top danger zones. According to the government-issued earthquake hazard map that Japan updates annually, the northeastern coast of Japan—where Fukushima is located—was thought to be among the least likely places in the country to experience a major earthquake. Some experts think that the Japanese government should de-emphasize the forecast because it pressures researchers to draw conclusions from the data that they gather (Harlan 2012).

## LITERATURE CITED

Diep. F. 2011. Fast facts about the Japan earthquake and tsunami, *Scientific American*, March 14th.

Fisher, J. 2011. Another earthquake shakes Japan. Fukushima evacuated; a nuclear timeline. *The Christian Science Monitor*, March 15.

Guo, H.D., Y. Chen, Q. Feng, Q.Z. Lin and F. Wang. 2011. Assessment of damage to buildings and farms during the 2011 M 9.0 earthquake and tsunami in Japan from remote sensing data, *Chinese Science Bulletin*. **56**:2138–2144.

Harlan, C. 2012. Temblor forecast rattles Japan, *Washington Post (newspaper)*, February 25th, A 8.

Harlan, C. 2012a. Japan: One year later. At a crossroads, *Washington Post (newspaper)*, March 11th, A 1, A 10.

ICRP (International Commission on Radiological Protection) 1991a. 1990 recommendations of the international commission on radiological protection, *ICRP publication 60, Annals of the ICRP.* **21 1-(3)**:1–201.

Kuczera, B. 2011. The severe earthquake in Tohoku in Japan and the impact on the Fukushima-Daiichi nuclear power station, *International Journal of Nuclear Power.* **56**:234–241.

Maynard, W.B. 2012. When the Big Muddy ran backward, *Washington Post (newspaper)*, January 31st, E 1 and E 6.

Miyagi, T., D. Higaki, H. Yagi, S. Doshida, N. Chiba, J. Umemura and G. Satch. 2011. Reconnaissance report on landslide disasters in northeast Japan following the M 9 Taihoku earthquake, *Landslides*. **8**:339–342.

National Geographic. 2005. Tsunami killer wave, *DVD video 1029*, 55 min.

Obe, M. 2012a. Japan reviews disaster plan amid new quake concerns, *Wall Street Journal (newspaper)*, January 24th, A14.

Osnos, E. 2011. Letter from Fukushima. The fallout, *The New Yorker*, October 17, pp. 46–61.

Smith, R. 2012. New risks for nuclear plants, *Wall Street Journal (newspaper)*, February 1st, A.3.

Smith, R. 2012a. Industry alters designs in an effort to make future plants safer, *Wall Street Journal (newspaper)*, March 9th, A 10.

Smith, R. 2012b. Cheap natural gas unplugs U.S. nuclear-power revival, *Wall Street Journal (newspaper)*, March 16th, A1, A10.

Suzuki, W., S. Aoi, H. Sekiguchi and T. Kunugi. 2011. Rupture process of the 2011 Toku-Oki mega-thrust earthquake (M9.0) inverted from strong-motion data, *Geophysical Research Letters*, 36. LOOG16, doi:10.1029/2011GLO49136.

Tabuchi, H. 2012. Independent panel challenges Japan's account of disaster at nuclear plant, *New York Times (newspaper)*, January 16th, B3.

Wikipedia. 2011. Timeline of the Fukushima Daiichi nuclear disaster. 21 p.

Chapter 3

# The Tsunami

## 3.1 General

Most large tsunamis arise around the rim of the Pacific Ocean and the Indian Ocean, along seafloor faults called subduction zones, where colliding tectonic plates trigger large earthquakes (Folger 2012). Waves spread in opposite directions from the fault and within minutes crash onto nearby land, as in Japan in 2011; within hours crossing the ocean and in this case sweeping a California man out to sea. Landslides and volcanic eruptions can launch tsunamis in lakes and rivers. A warning system consisting of a network of 53 detector buoys, most in the Pacific Ocean, now tracks the movement of tsunamis, alerting people on distant shores. Major tsunamis occur at the rate of once a year globally, with records of thousands killed in Chile, Portugal and Sumatra—with one predicted for the Cascadia subduction zone extending along the Pacific coast of North America (Folger 2012).

The earthquake and tsunami that struck Japan in March 2011 created an estimated 25 million tons of debris, large amounts of which washed into the ocean (Grossman 2012). In September 2011, a Russian ship found a Japanese fishing vessel, a refrigerator and other household items near Midway Atoll. In December, Japanese fishing floats washed up in Washington State and in Vancouver, British Columbia (Grossman 2012). A five-year projection of debris from the 2011 earthquake and tsunami shows that the debris is likely to spread via the North Pacific Subtropical spiral of currents (Berlin 2012). Based on satellite data and drifting behavior of scientific buoys, the model shows debris reaching Midway Island in 2012, Hawaii in 2013, the North American coast in 2014 and the North Pacific Garbage Patch (midway between Hawaii and California) in 2015 (Berlin 2012).

Of the estimated 5 million tons of tsunami-related debris, about 3.5 million tons sank off the Japanese coast (Millman 2012). The largest piece of debris recorded to date is a 65,000 kg slab of reinforced cement from Japan that washed ashore at Agate Beach, Oregon in early June, 2012. It measured about 23 m in length, 6 m wide and 2.5 m high, being one of four floating docks built for a Japanese fishing fleet in Misawa, Japan. Oregon state workers used torches to kill invasive sea creatures from the dock. The dock was eventually dismantled for $84,155 (Millman 2012).

## 3.2 Japan

Unlike earthquakes, which can occur daily in Japan, tsunamis (harbor waves in Japanese) often skip a generation, giving them a hidden and unpredictable power (Mockett 2012). The coastline of northeast Japan is dotted with stones hundreds of years old inscribed with warnings of past tsunamis (Mockett 2012).

The earthquake of March 11th caused a tsunami because the seafloor was suddenly forced to move vertically due to subduction (Diep 2011). The process of subduction wherein two tectonic plates push against each other then separate with energy release resulted in this case of a 15-meter (49 feet) high wave that can travel at speeds up to 800 km/hour, giving residents of Sendai only 8 to 10 minutes of lead time (Diep 2011). On March 11, 2011, the 15-m wave overtopped the 5.7 m seawall surrounding the Fukushima facility, inundating and destroying the backup diesel generators—all but one of which were housed underground—and washing away the fuel tanks (Wikipedia 2011). With the loss of electric power supply, most of the emergency core cooling system—including the low-pressure core spray, the residual heat removal, the low-pressure coolant injection system main pump and the automatic depression systems—all failed (Wikipedia 2011). At this point, TEPCO notified authorities of a "First level emergency" as required by law (Wikipedia 2011c).

There were many tsunamis recorded by October 21st, 2011 according to the Japan Meteorological Society, with heights ranging from 10 to 38 meters (Takahashi et al., 2012). According to the National

Police Agency of Japan, these tsunamis completely destroyed more than 90 percent of the dwellings in the towns that they struck, killed 15,828 people—(with 3,760 still reported missing), destroyed 302,066 homes and 3,559 roads and displaced as many as 470,000 people (Takahashi et al., 2012). The March 2011 tsunami extended over a 340-km long section of the coast in northeastern Honshu Island, as determined by Advanced Land Observing Satellite (ALOS) and aerial photographs (Rao and Lin 2011). It inundated all property inland up to 5.5 km from the coast in low-lying areas and up to 15 km in a narrow bay at the mouth of a large river. High run-ups up to 35 m occurred in coastline areas close to the epicenter (Rao and Lin 2011).

The Fukushima tsunami—at its maximum run-up height of 39 m (Mimura et al., 2011)—destroyed many coastal cities (Takahashi et al., 2011), penetrating more than 3.6 km inland (Folger 2012). By early May, over 24,000 people were reported as dead or missing (Mimura et al., 2011) and hundreds of thousands homeless (Folger 2012). As a result the Study Group on Guidelines for the First Steps and Emergency Triage to Manage Elderly Evacuees quickly established guidelines enabling non-medical care providers (including volunteers, helpers, family members taking care of elderly relatives), public health nurses and certified social workers to rapidly detect illnesses in elderly evacuees. More than 20,000 booklets were distributed to care providers in Fukushima, Iwate and Miyagi prefectures with the goal of reducing susceptibility to disaster-related illnesses (infectious diseases, exacerbation of underlying illnesses and mental stress) and deaths in elderly evacuees (Takahashi et al., 2011)

In a report to the International Atomic Energy Agency (IAEA 2011) the Japanese government stated that the Fukushima disaster was caused by the tsunami—not by the earthquake—resulting in power loss to the cooling systems, producing three core meltdowns (Noggerath et al., 2011). The Fukushima reactors were designed in the 1960s according to the best scientific knowledge available. But between the 1970s and 2011, new scientific knowledge emerged about the likelihood of a large earthquake and resulting tsunami. This advice was ignored by both TEPCO (Tokyo Electric Power Company) and government regulators. The regulatory authorities failed to properly review the tsunami countermeasures in accordance with IAEA (2011) guidelines and continued to allow the plant to operate without sufficient countermeasures, despite clear warnings from members of a government advisory committee. The lack of independence of government regulators appears to have contributed to this inaction.

Japan's seismological agencies at that time seemed to focus on earthquake hazards in the Tokai district, located between Tokyo and Nagoya, while downplaying earthquake hazards elsewhere in Japan, thus missing opportunities to prevent the Fukushima calamity (Noggerath et al., 2011).

After the diesel generators in the turbine buildings failed, emergency power for control systems was supplied by batteries designed to last about eight hours. Additional batteries and mobile generators were dispatched to the site; owing to poor road conditions, these arrived almost six hours after the tsunami struck.. Due to flooding, all attempts to connect the portable generating equipment to power water pumps were discontinued (Wikipedia 2011c). A nuclear reactor generates heat by splitting atoms, typically uranium, in a chain reaction (Wikipedia 2011c). The reactor continues to generate heat after the chain reaction is stopped owing to radioactive decay of unstable isotopes. This decay of unstable isotopes and the decay heat can't be stopped. Immediately after shutdown, decay heat amounts to about 6% of full thermal heat production of the reactor. The decay heat in the reactor core decreases over several days before reaching cold shutdown levels. Nuclear fuel rods that have reached cold shutdown temperatures usually require several years of water cooling in a spent fuel pool before decay heat production reduces to the point that they can be safely transferred to dry storage. To safely remove the decay heat, reactor operators need to continually circulate cooling water over fuel rods in the reactor core and spent fuel pond (Wikipedia 2011c).

The kinetic energy of the tsunami is determined by frictional release on the offshore slope to seabed, nearshore sand and sediment supply and coast linearity of embayments that constrict wave energy (Mahaney and Dohm 2011). At Fukushima, for example, with gentle offshore gradients and relatively linear coastlines, the kinetic energy of the tsunami is decreased by friction with the seabed and radiated outward along the coast, thus producing reduced flooding of the affected coast. In contrast, steep offshore gradients—typical of Sendai—reduce friction to mere milliseconds of wave impact and would produce higher-energy waves of greater magnitude, as attested by greater penetration inland with considerable loss of life and property (Mahaney and Dohm 2011). This tsunami destroyed 24% of the buildings and 12% of the farms in neighboring Miyagi Prefecture, with lesser damage reported to buildings and farms in Fukushima Prefecture (Guo et al., 2011). On June 24, 2009, two senior Japanese seismologists warned TEPCO that the Fukushima reactors were acutely vulnerable to tsunamis (Osnos 2011).

The March 11 tsunami was history's most expensive natural disaster with losses estimated at 300 billion dollars. After the tsunami, TEPCO barred rank and file employees from speaking publicly, a ban that was still in effect on October 17th, more than 7 months later (Osnos 2011).

## LITERATURE CITED


Berlin, J. 2012. Tsunami debris path, *National Geographic.* **22(1)**:26.

Diep. F. 2011. Fast facts about the Japan earthquake and tsunami, *Scientific American,* March 14th.

Folger, T. 2012. The calm before the wave, *National Geographic.* February. **221(2)**:54-77.

Grossman, E. 2012. Remains of the day, *Scientific American,* March, 11.

Guo, H.D., Y. Chen, Q. Feng, Q.Z. Lin and F. Wang.2011. Assessment of damage to buildings and farms during the 2011 M 9.0 earthquake and tsunami in Japan from remote sensing data, *Chinese Science Bulletin.* **56**:2138–2144.

IAEA (International Atomic Energy Agency). 2011. Fukushima nuclear accident update log, iaea.org/...tsunamiupdate01.html

Mahaney, W.C. and J.M. Dohm. 2011. The 2011 Japanese 9.0 magnitude earthquake: Test of a kinetic energy wave model using coastal configuration and offshore gradient of Earth and beyond, *Sedimentary Geology.* **239(1-2)**:80–86.

Millman, J. 2012. Tsunami relic puts a beach on the map, *The Wall Street Journal (newspaper),* June 21st, A6.

Mockett, M.M. 2012. A time to run, *National Geographic.* February, **221(2)**:78–79.

Noggerath, J., R,J. Geller and V.K. Gusiakov. 2011. Fukushima: The myth of safety, the reality of geoscience, *Bulletin of the Atomic Scientists,* **67(5)**:37–46.

Osnos, E. 2011. Letter from Fukushima. The fallout. *The New Yorker,* October 17, pp. 46–61.

Rao, G. and A. Lin. 2011. Distribution of inundation by the great tsunami of the 2011 Mw 9.0 earthquake off the Pacific coast of Tohoku (Japan), as revealed by ALOS imagery data, *International Journal of Remote Sensing,* **32(22)**:7073–7086.

Takahashi, T., K. Iijima. M. Kuzuya, H. Hattori, K. Yokono and S. Morimoto. 2011. Guidelines for non-medical care providers to manage the first steps of emergency triage of elderly evacuees, *Geriatrics & Gerontology International.* **11(4)**:383–394.

Takahashi, T., M. Goto, H. Yoshida, H. Sumino and H. Matsui. 2012. Infectious diseases after the 2011 Great East Japan earthquake, *Journal of Experimental and Clinical Medicine.* **4(1)**:20–23.

Wikipedia. 2011. Timeline of the Fukushima Daiichi nuclear disaster, 21 p.

Wikipedia. 2011c. Fukushima Daiichi nuclear disaster, 54 p.

CHAPTER 4

# Fukushima Nuclear Reactor

## 4.1 Safety History

The Fukushima Daiichi nuclear power complex was central to a falsified-records scandal that led to the departure of a number of senior executives of TEPCO as well as disclosures of previously unreported problems at the plant (Wikipedia 2011c). In 2002, TEPCO admitted it had falsified safety records at the No. 1 reactor. Because of the scandal and a fuel leak at Fukushima, the company shut down all of its 17 reactors. A power board distributing electricity to a reactors' temperature control valves was not examined for 11 years and inspections did not cover devices relating to cooling systems such as water pump motors and diesel generators. In addition to Japanese concerns about seismic activity, the International Atomic Energy Agency (2011) expressed concern about earthquake vulnerability and warned, in 2008, that a strong earthquake of 7.0 or greater could pose a serious problem. In March 2006, the Japanese government opposed a court order to close a nuclear plant in western Japan over its ability to withstand an earthquake. Japan's Nuclear and Industrial Safety Agency (NISA) believed that it was safe and that all safety analyses were appropriately conducted. On October 2nd, 2011, on request of the Japan Broadcasting Corporation, the Japanese government released a report from TEPCO to NISA revealing that TEPCO was aware of the possibility that the plant could be hit by a tsunami wave far higher than the 5.7 meters which the plant was designed to withstand. Simulations done in 2008, based on the 1896 earthquake in this area, made it clear that waves between 8.4 and 10.2 meters could overflow the plant. Three years later, the report was sent to NISA where it arrived on March 7th, just 4 days before the tsunami. The issue was shelved until October 2012. TEPCO officials said the company did not feel the need to take prompt action on the estimates, which were still tentative calculations in the research stage, although NISA said that these results should have been made public and that the firm should have taken measures immediately (Wikipedia 2011c).

New designs under development address safety problems in Fukushima reactors (Clery 2011). The Fukushima reactors are based on U.S. commercial plant reactors from the 1960s and are technologically inferior to the newer machines that incorporate improvements developed over the past 50 years, the so-called Generation III designs. To design a safe reactor is to first shut the fission reaction down and second to cool the fuel. As at Three Mile Island, the three operating reactors at Fukushima (three were offline for refueling or repairs) executed the first of these tasks automatically following the earthquake. In boiling water reactors (BWRs) of the type used at Fukushima, neutron-absorbing control rods are pushed upward from below the core to between the fuel rods, killing the fission chain reaction—although this still requires power. In modern reactors, however, the rods are held above the core with electromagnets so that a power outage will release them and gravity will do the rest. But stopping the chain reaction doesn't neutralize heat production in the core. Radioactive decay of the fuel and fission products keeps generating up to 7% of the thermal capacity of the reactor and cooling water must be actively pumped through the core to prevent overheating, as was true for Fukushima. The new approach to power supply is to have more than one connection to the grid, an oversupply of diesel generators at different places and batteries as a backup. All Generation III designs emphasize moving coolant with gravity and convection, driving pumps with steam, activating valves with DC battery power or no power at all and keeping humans out of the loop. One Gen III design dispenses with pumps altogether to circulate the coolant. Other safety measures include automatic pressure-release valves, a suppression tank to which steam can escape when pressure gets high and water reservoirs above the reactor to top up the coolant via gravity if the level drops too low (Clery 2011).

The AP 1000 series by Westinghouse has reduced the piping, cables and valves in order to reduce the possibility of problems (Clery 2011).

China is building the first four of the Gen III which uses convection to shift heat from the reactor vessel into a massive tank of nearby cooling water. Decay heat is thus managed naturally without the need for diesel generators. Modern reactor designs don't keep fuel in an elevated pool above the reactor like BWRs but store it in a separate earthquake-proof building. But the spent fuel problem remains a global problem. The days may be numbered for old BWRs, like those of Fukushima, but nuclear will remain firmly in the mix of future energy sources (Clery 2011).

It is not possible to fully anticipate and account for all human errors and unusual or "beyond design" events (Chokshi 2011). The design of nuclear power plants utilizes probability risk assessments to ensure that accidents are unlikely. The probability of core damage in new reactors may be theoretically less than 1 in 1 million reactor years. However, the real-life frequency of a nuclear accident may be as high 1 in 3,000 reactor years based on the 15,000 reactor years of operation of nuclear power and the known core damage in five reactors (Three Mile Island, Chernobyl and three reactors in Fukushima). For the currently existing 400 plus reactors world-wide, the real-life frequency (without significant upgrade) suggests the possibility of an accident once every eight years (Chokshi 2011). The situation in Japan is different with 40 of the nation's 54 reactors currently offline (Obe 2012a). By May, 2012, these last five will be required to close for maintenance (Obe 2012a).

A nuclear accident like Fukushima is unlikely elsewhere, provided that an effective nuclear regulatory system is in place (Wang and Chen 2011). Authors believe that the nuclear safety regulatory system should be upgraded to ensure the safety of the 433 operational reactors and the 65 currently under construction worldwide. Specifically, each country should have an independent agency for reactor safety that is legitimate, credible and authoritative; legislation should be enacted for nuclear personnel conflict-of-interest restrictions, even after they leave a nuclear agency; and finally, international peer-review to ensure compliance with nuclear safety standards (Wang and Chen 2011).

## 4.2 Damage from March 11th Quake and Tsunami

The Fukushima Daiichi nuclear power facility in the Futaba district of the Fukushima Prefecture in Japan was severely damaged by the earthquake and ensuing tsunami that struck off the northern coast of Honshu on March 3, 2011 (Dauer et al., 2011). The resulting structural damage disabled the reactor's cooling systems and led to significant, ongoing, environmental releases of radioactivity, triggering a mandatory evacuation of a large area surrounding the plant (Dauer et al., 2011). In 1990, the U.S. Nuclear Regulatory Commission (NRC) ranked the failure of the emergency electricity generators and subsequent failure of the cooling systems of nuclear plants in seismically very active regions one of the most likely risks (Wikipedia 2011b). The Japanese Nuclear and Industrial Safety Agency (NISA) cited this report in 2004 but apparently TEPCO did not react to these warnings and did not respond with any measures (Wikipedia 2011b).

The core of each operating reactor holds at least 25,000-19 m fuel rods, each filled with pellets of enriched uranium (Osnos 2011). The heat produced by these pellets was used to boil water for steam that drove turbines that generated electricity. Radioactive isotopes that were produced during this process were normally contained safely. But with power and emergency systems down, there was no way to cool the hot fuel. Eventually the coolant water boiled away causing reactor fuels to melt, eating through the enclosed shells. At the plant, the first tsunami wave of 5.6 m in height arrived at 3:27 PM on March 11 but did not overtop the 5.7 m concrete seawall surrounding the reactors. After the first wave had receded, survivors ventured down to the water's edge to see who could be saved, only to be swept away by a much larger second wave 15 m (49 feet) high. In all, 200,000 people died or disappeared along a 600-km stretch of the Japanese coast. The second wave arrived only 8 minutes later and advanced towards the reactors. Two minutes after the water arrived, the plant's main control rooms began to lose electric power without a constant source of coolant, the nuclear fuel rods in the heart of the three active reactors would eventually boil away the water that prevents meltdown. Seawater swamped the plant's emergency diesel generators stored on the ground floor and basements. These were destroyed and 2 workers who had been sent underground to check for leaks were drowned (Osnos 2011).

On March 11th, fuel elements in the core of reactor 1 were exposed above water just 4 hours after the earthquake and had fully melted after 16 hours (Wikipedia 2011). On March 12th, at 3:36 PM there was a massive explosion in the outer structure of unit 1, with the collapse of the concrete building surrounding the steel reactor vessel. All residents within 10 km were told to evacuate. By 9:40 PM. The evacuation zone around Fukushima 1 was extended to 20 km and the evacuation zone around Fukushima 2 was extended to 10 km. On March 13th, reactors 1 and 3 were vented to the atmosphere to release excess pressure and then refilled with water and boric acid for cooling and to inhibit further nuclear reaction. The Japan Atomic Energy Agency announced that it was treating the situation at reactor 1 at Level 4 (an accident with local consequences) on the International Nuclear and Radiological Event Scale (= INES) (Wikipedia 2011).

Tokyo Electric Co. crews pumped seawater into a third reactor to halt a meltdown of its fuel assemblies hours after a second explosion leaked hydrogen gas and rocked another reactor building at the site (Behr and Wire 2011). Reports from government and industry officials in Japan initially minimized the threat. The U.S. Nuclear Energy Institute stated, erroneously, that the Unit 1 reactor core still had a sufficient amount of water for cooling "with no danger of the nuclear fuel being exposed" (Behr and Wire 2011). It is noteworthy that the agency charged with policing Japanese power plants—the Nuclear and Industrial Safety Agency—is part of the same Japanese ministry in charge of promoting power plants; incidentally, after the Chernobyl accident in the Ukraine in 1986, Japan—unlike many countries—began construction of 5 additional reactors (Osnos 2011).

The reactor in building number 1 exploded, hurling chunks of concrete that injured 5 workers (Wikipedia 2011). On the third day after the tsunami, reactor 3 exploded and on day 4, reactor 4 exploded; all explosions released massive doses of radiation from the isotope storage facilities (Wikipedia 2011). At 7:03 PM on March 11, a nuclear emergency was declared and all people within 3 km of the plant were ordered to evacuate; eventually, the number of nuclear refugees exceeded 80,000. Fuel that had already melted into a heap at the reactor bottom could theoretically melt through the steel pressure vessel and react with the concrete below, releasing radioactive materials such as radiostrontium and radiotechnetium. The impact might reach the outskirts of Tokyo, with a population of 35 million. Nevertheless, that evening a spokesman for the Japanese government stated "there is no radiation leak nor will there be a leak". The world's worst nuclear accident since Chernobyl in 1986 was allegedly man-made, the consequences of failures laid bare by how far Japan's political and technological rigor had drifted (Osnos 2011).

A handful of dedicated plant workers remained on the site, implementing emergency cooling measures at the overheating reactors (Biello 2011; Matson 2011). The extent to which their health has been endangered may not become apparent for years. Unlike the 1986 Chernobyl disaster in the Ukraine, in which dozens of cases of fatal radiation poisoning among plant workers and in which thousands of thyroid cancer were diagnosed in the ensuing years (Eisler 1995, 2000,

2003, 2007), radiation releases at Fukushima have been limited (Matson 2011); further, the Fukushima reactors were better designed than the failed Chernobyl reactor (Matson 2011).

Three previous civilian meltdowns are known: Three Mile Island; Saint-Laurent Nuclear Power Plant (France); and Chernobyl (Wikipedia 2011c). At first, TEPCO denied that fuel cores were melted and considered removing all staff members from the plant and leaving it abandoned. This option was rejected by the Japanese government on the grounds that radiation leaks dozens of times greater than that released at Chernobyl would result in the evacuation of Tokyo and could compromise the very existence of the Japanese nation (Wikipedia 2011c).

On April 10th, TEPCO began using remote-controlled unmanned heavy equipment to remove debris from around nuclear reactors 2–4 (Wikipedia 2011c). When the monsoon season began in June, a light fabric cover was used to protect the damaged reactor buildings from storms and heavy rainfall. On August 1st, high radiation levels were found outside buildings 1 and in an exhaust-pipe. On August 16th, TEPCO installed devices in the spent fuel pools of reactors 2, 3 and 4 which used special membranes and electricity to desalinate the seawater (to prevent damage to the steel pipes and the pool walls). TEPCO also outlined the steps needed for long-term core removal, but admitted that existing technologies were unsatisfactory and that new ones must be invented involving robots and remote controlled machinery to remove debris. In the second step, the primary containment vessels will be inspected, water leakage paths identified, then sealed. Once the containment vessels are air and water-tight, cooling systems will be installed using heat exchangers and various water circulation devices. The final step calls for the containment vessel and pressure vessels to be opened from the top (as in normal refueling) and for special robotic arms to descend and remove the melted fuel from the bottom of the pressure vessels as well as any fuel that leaked to the floor of the outer containment vessels (Wikipedia 2011c). The private insurance industry will not be significantly affected by the accidents at the Fukushima nuclear power plant because "coverage for nuclear facilities in Japan excludes earthquake shock, fire following earthquake and tsunami, for both physical damage and liability"(Wikipedia 2011c).

On November 8th, the ground around the reactors remained littered with mangled trucks, twisted metal beams and broken building frames (Fackler 2011). Japan faces decades of budget-draining cleanup before the surrounding countryside can become habitable again.

The radiation level at the Fukushima visitor center was 13 tunes the recommended maximum annual dosage for civilians. In other spots radiation was sufficiently elevated to reach the annual recommended maximum dosage in only 3 hours. But reactor 1 had been capped with a superstructure to trap radioactive materials. And American, French and Japanese scientists had constructed a massive system for decontaminating water. The water is part of a new cooling system that can reduce temperatures in the damaged reactor cores below boiling, a necessary first step in cold shutdown. Also present was a field full of newly-constructed four-story tall silver tanks that housed much of the 90,000 tons of contaminated water that had been dumped on the reactors. TEPCO spokesman stated that the reactors were stabilized although danger remained. Among remaining challenges was the condition of fuel within No. 1 and 3 reactors, whose cores appear to have melted through the inner containment vessels. Cold shutdown is an indication that the accident phase is over, but the next phase of cleanup will take more than 20 years (Fackler 2011).

After explosions, massive radioisotope releases and immersion in sea water, TEPCO announced that it will scrap its Fukushima reactors (Brumfiel 2012a). When decommissioning the facility, workers may encounter warped, split and partially melted fuel rods. They may discover that molten fuel flowed into the outer containment vessel or that nuclear chain reactions are still happening inside the fuel. Particularly worrisome is circumstantial evidence of the presence of chorine-38, an isotope with a half-life of just 37 minutes, which forms when natural chlorine-37 is hit by neutrons from fission, suggesting fuel has clumped into sufficiently large chunks to briefly restart nuclear reactions. Such bursts put workers at extreme risk of radiation exposure during cleanup. The confusion recalls the weeks that followed the partial meltdown of a reactor at Three Mile Island near Harrisburg, Pennsylvania in 1979. It took 14 years to clear most of the fuel from the reactor at Three Mile Island, suggesting that the decommissioning of Fukushima will probably take longer (Brumfiel 2011a).

In early 2012, Japanese authorities introduced stress tests for all nuclear reactors in the country in an effort to ease public worries over nuclear-power safety, but the tests were considered inadequate by several experts (Iwata 2012a). Specifically, only a very limited aspect of comprehensive reactor safety was considered. All tests were due to be completed by the end of 2011, but now completion by the end of summer is also doubtful. In any event, public fears have not been assuaged (Iwata 2012a).

## 4.3 Evacuees

On March 12th, 2011, the government ordered total evacuation within a 20-km radius around the nuclear power plant, although most residents had already left; however, about 1,500 patients still remained in hospitals or nursing facilities (Tanigawa et al., 2011). On March 14th, these patients were transported to a Health Care Center 24 km north of the damaged reactors by buses, police vehicles and Japan Self Defense Forces transport. Many patients had to wait for more than 24 hours in the transport vehicles in the cold weather without food or water; at least 21 elderly patients died from hypothermia or dehydration. An additional 1,700 patients remained in the area and the government decided to move everyone from the zone by March 20th. Initially, a value of 13,000 counts per minute, as measured by a Geiger-Muller counter, was used to indicate the need for evacuee decontamination. This was raised to 100,000 counts per minute because of shortages of water and other vital supplies, limited decontamination facilities, very low temperatures and the need to cope with thousands of evacuees. More than 20,000 people had been screened at 13 Centers by March 16th (Tanigawa et al., 2011).

Many displaced people moved into shelters or temporary homes supplied by the government because of disruptions to community utility services and health risks associated with nuclear power plant malfunctions in Fukushima (Takahashi et al., 2012). In general, they were exposed to cold, unhygienic conditions and malnutrition owing to insufficient food provision and lack of running water. Elevated frequencies of infectious diseases were recorded among the elderly and infants, including respiratory tract infections, tetanus, measles and food poisoning (Takahashi et al., 2012).

With concern about the radiological emergency growing, one of Japan's largest hospital and healthcare groups issued a request for assistance to a U.S.-based international disaster relief organization (Becker 2011). After consultation with the Japanese, a special Radiation Emergency Assistance Mission was assembled. In April 2011 the mission arrived in Japan with several aims: (1) to rapidly assess the situation on the ground, (2) to exchange information, experiences and insights with Japanese colleagues and (3) provide radiological information and practical refresher training to Japanese healthcare professionals and first responders. The mission broke new ground. Whereas government-to-government assistance requests and aid missions occur regularly after disasters, a non-governmental organization (NGO) to NGO arranged mission with a specific focus on radiological issues is unusual (Becker 2011).

In late January 2012, government officials announced that Namie—6 km from the Fukushima reactors—would likely remain evacuated for "several years" or longer owing to more than 25,000 Bq/kg of cesium-137 in its soil (Hayashi 2012). This coastal village of 21,000—best known for its fishing port, scenic gorge and pottery—has survived for hundreds of years, with some businesses operated by the same family for 25 generations. A similar situation exists in other communities forced to evacuate. In Namie, where most people had lived for at least 20 years, some (2,800) relocated to Nihomatsu, 48 km inland but most (52%) of the younger residents age 34 or younger said that they wouldn't return under any circumstance vs. 17% for those age 65 to 79. Nearly half of Namie's residents said that they have moved at least three times since the accident. Before March 11th, 2011. Namie's school system had 1,710 children in its elementary and middle schools vs. a current enrollment of 82. Preschoolers wear dosimeters around their necks. The need for Namie's citizens to find permanent new homes and jobs (current unemployment rate is 50%) accelerates and this is typical for other evacuated communities (Hayashi 2012).

By March 11th, 2012—one year after the incident—nothing seems to have been resolved in Ishinomaki (Harlan 2012a). Once a town of 153,000, one of every three homes there had been damaged or destroyed by the wave. The tsunami-related debris has been removed and reconstruction remains at a starting point, depending on whether people remain on site. A full recovery, if possible, will take at least a decade. Workers at the City Hall were trying to attract clean-energy projects and were offering tax incentives for businesses that relocated there. To date, about 6,000 have fled and 7,000 live in temporary housing. Disaster survivors rushed to rebuild in spots that hadn't been leveled, driving up prices sharply and creating deep rifts in the community. Despite massive volunteer help (240,000 in the past year) with debris removal, fixing damaged homes, filling in cracked walls, repairing shutters, etc., the city can only recover partially. There is a 10-year reconstruction plan aimed at restoring old factories and ports, while also attracting renewable energy products. The first 3 years of the plan will focus on infrastructure and importing mass amounts of soil in order to raise the coastal land (Harlan 2012a).

## LITERATURE CITED

Becker, S.M. 2011. Learning from the 2011 Great East Japan disaster: insights from a special radiological emergency assistance mission, *Biosecurity and Bioterrorism: Biodefense Strategy, Practice and Science.* **9(4):**394–404.

Behr, P. and C. Wire. 2011. Will Fukushima disaster spell the end for a U.S. nuclear revival? *Scientific American,* March 14th.

Biello, D. 2011. Workers battle Fukushima nuclear crisis at personal risk, *Scientific American,* March 16th.

Bird, W.A. 2011. Fukushima health study launched, *Environmental Health Perspectives.* **119(10):**a428-a429.

Blustein, P. 2012. A wake-up call Japan ignored, *Washington Post (newspaper),* March 11th, B 3.

Brumfiel, G. 2011a. Japan's long road ahead, *Nature.* **472(7342):**146–147.

Chokshi, A.H. 2011. Fukushima: India at crossroads, *Current Science.* **100(11):**1603.

Clery, D. 2011. Current designs address safety problems in Fukushima reactors, *Science.* **331(6024):**1506.

Dauer, L., P. Zanzonico, R.M. Tuttle, D.M. Quinn and H.W. Strauss. 2011. The Japanese tsunami and resulting nuclear emergency at the Fukushima Daiichi power facility: Technical, Radiologic and Response Perspectives, *Journal of Nuclear Medicine.* **52:**1423–1432.

Eisler, R. 1995. Ecological and toxicological aspects of the partial meltdown of the Chernobyl nuclear plant reactor. In: D.E. Hoffman, B.A. Rattner, G.A.Burton Jr. and A.J. Cairns (eds). Handbook of Ecotoxicology, Lewis Publishers, Boca Raton, Florida, pp. 459–564.

Eisler, R. 2000. Radiation. In Handbook of Chemical Risk Assessment, Volume 3. Lewis Publishers, Boca Raton, Florida, pp. 1707–1828.

Eisler, R. 2003. The Chernobyl nuclear power plant reactor accident: Ecotoxicological update. In: D.J. Hoffman, B.A. Rattner, G.A. Burton Jr. and A.J. Cairns (eds.). Handbook of Ecotoxicology, Second Edition, Lewis Publishers, Boca Raton, Florida, pp. 703–736.

Eisler, R. 2007. Eisler's Encyclopedia of Environmentally Hazardous priority Chemicals. Chapter 27, Radiation, Elsevier, Amsterdam. pp. 677–736.

Fackler, M. 2012. Japanese struggle to protect their food supply, *New York Times (newspaper),* January 22nd, A6.

Harlan, C. 2012a. Japan: One year later. At a crossroads, *Washington Post (newspaper),* March 11th, A 1, A 10.

Hayashi, Y. 2012. Displaced Japanese town tries to stay intact, *Wall Street Journal (newspaper),* March 3-4, A 8.

IAEA (International Atomic Energy Agency). 2011. Fukushima nuclear accident update log, iaea.org/...tsunamiupdate01.html

Iwata, M. 2012a. Japan nuclear stress tests fail to assuage public fears, *Wall Street Journal (newspaper).* March 3-4, A 8.

Matson, J. 2011. Fast facts about radiation from the Fukushima Daiichi nuclear reactors, *Scientific American,* March 16th.

Obe, M. 2012a. Japan reviews disaster plan amid new quake concerns, *Wall Street Journal (newspaper),* January 24th, A14.

Osnos, E. 2011. Letter from Fukushima. The fallout. *The New Yorker,* October 17, pp. 46-61.

Takahashi, T., M. Goto, H. Yoshida, H. Sumino and H. Matsui. 2012. Infectious diseases after the 2011 Great East Japan earthquake, *Journal of Experimental and Clinical Medicine.* **4(1)**:20–23.

Tanigawa, K., Y. Hosoi, N. Hirohashi, Y. Iwasaki and K. Kamiya, 2011. Evacuation from the restricted zone of the damaged Fukushima nuclear power plant: facing with the reality, *Resuscitation.* **82(9)**:1248.

Wang, Q. and X. Chen. 2011. Nuclear accident like Fukushima unlikely in the rest of the world?, *Environmental Science & Technology.* **45(23)**:9831–9832.

Wikipedia. 2011. Timeline of the Fukushima Daiichi nuclear disaster, 21 p.

Wikipedia. 2011b. Fukushima Daiichi nuclear power plant, 7 p.

Wikipedia. 2011c. Fukushima Daiichi nuclear disaster, 54 p.

Chapter 5

# Radiation Releases

## 5.1 General

In the first hours of the March 11th accident, workers rushed to flood three damaged reactors with seawater to prevent a catastrophic meltdown (Brumfiel and Cyranoski 2011). Three months later, water was still being pumped into the cores and has since become the biggest obstacle to cleaning up the site. Residual nuclear decay in the three reactors—which all suffered total meltdowns—means that they will need cooling for many months. TEPCO switched to using fresh water two weeks after the accident as the salt water corroded the stainless steel reactor vessels. Eventually, TEPCO dumped more than 10,000 tons of low-level contaminated water into the Pacific Ocean and admitted that several hundred tons of highly contaminated water also leaked out, exposing marine life to large doses of radiation (Brumfiel and Cyranoski 2011).

By March 15th, the Japanese government publicly maintained that the nuclear risk was manageable; however, U.S. authorities disagreed and developed plans to evacuate the estimated 100,000 American citizens from Tokyo (Osnos 2011). On March 16th, a few days after the accident, the U.S. Nuclear Regulatory Commission advised Americans in the region to evacuate out to 50 miles (80 km) from the Daiichi nuclear complex; if the Japanese government had made the same recommendation to its citizens, it would have resulted in the early evacuation of about 2 million people, instead of 130,000 (Hippel

2011). But the U.S. decision was based on faulty data: the belief that no water remained in a pool at reactor 4 that was storing spent fuel rods (Landers 2012). In fact, the number 4 pool was nearly undamaged and the spent fuel rods remained covered with water throughout (Landers 2012), once again demonstrating that faulty communications occur regularly during crises.

Four major isotopes of possible health concern were released from Fukushima: iodine-131, cesium-137, strontium-90 and plutonium-239 (Wikipedia 2011a). Iodine-131, with a short half-life of about 8 days, is rapidly absorbed by the thyroid gland and persons exposed to releases of iodine-131 from any source have a higher risk for developing thyroid cancer, with children more vulnerable than adults. Increased risk for thyroid neoplasm remains elevated for at least 40 years after exposure. Potassium iodide tablets are known to prevent iodine-131 absorption, although the number of doses of stable potassium iodide available to the Japanese public was inadequate to meet the perceived needs for a Fukushima-type nuclear emergency. Cesium-137, with a half-life of 30 years, behaves like potassium and is taken up by cells throughout the body. Cesium-137 can cause acute radiation sickness and increase the risk for cancer owing to high-energy gamma radiation. Internal exposure to cesium-137 through ingestion or inhalation, allows radioactive material to be distributed in soft tissues, especially muscle, exposing these tissues to beta particles and gamma radiation and increasing cancer risk. Strontium-90, with a half-life of 30 years, behaves like calcium and tends to deposit in bone and bone marrow. About 25% of ingested strontium-90 is absorbed and deposited in the bone. Internal exposure to strontium-90 is linked to bone cancer, cancer of the soft tissue near the bone and leukemia. Risk of cancer increases with increased exposure to strontium-90. Plutonium-239 is particularly long-lived and toxic with a half-life of 24,000 years. It is present in spent fuel rods and in fuel of the Unit 3 reactor. It remains hazardous for tens of thousands of years, but long-term risk associated with plutonium-239 is highly dependent on the geochemistry of the particular site and in the case of Fukushima poses negligible risk to human health (Wikipedia 2011a).

On March 19th trace amounts of radioactivity were detected in Tokyo tap water; also, radiation levels exceeding legal limits were detected in milk and vegetables from the Fukushima area (Table 5.1; Table 5.2; Wikipedia 2011a, 2011c). On March 19th, upland soil 40 km distant from Fukushima contained 28,100 Bq/kg of cesium-137 and 300,000 Bq/kg of iodine-131. (See Table 5.1) One day later these same

**Table 5.1 New unitsa for use with radiation and radioactivity, measurements and conversions (International Commission on Radiological Protection 1977, 1991a; Hobbs and McClellan 1986; United Nations Scientific Committee on the effects of atomic radiation 1988).**

NEW UNITS

| *Variable* | *Old unit* | *New Unit* | *Old unit in terms of new unit* |
|---|---|---|---|
| Activity | Curie (Ci) = $3.7 \times 10^{10}$ disintegrations per second (dps) | Becquerel (Bq) = 1 dps | 1 Ci = $3.7 \times 10^{10}$ Bq |
| Exposure | Roentgen (R) = $2.58 \times 10^{-4}$ Coulombs/kg | Coulomb/kg (C/kg) | 1 R = $2.58 \times 10^{-4}$ C/kg |
| Absorbed dose | Rad= 100 erg/g | Grey (Gy) = 1 J/kg | 1 Rad = 0.01 Gy |
| Dose equivalent | Rem = damage effects of 1 R | Sievert (Sv) = 1 J/kg | 1 Rem = 0.01 Sv |

PREFIXES USED IN CONJUNCTION WITH STANDARD UNITS

| *Multiple* | *Prefix* | *Symbol* |
|---|---|---|
| $10^{12}$ | tera | T |
| $10^{9}$ | giga | G |
| $10^{6}$ | mega | M |
| $10^{3}$ | kilo | k |
| $10^{-2}$ | centi | c |
| $10^{-3}$ | milli | m |
| $10^{-6}$ | micro | *u* |
| $10^{-9}$ | nano | n |

CONVERSIONS

| | *Conversion Equivalence* |
|---|---|
| 1 curie = $3.7 \times 10^{10}$ dps<br>1 millicurie (mCi)<br>1 rad<br>1 rem | 1 becquerel = 1 disintegration per second<br>= 37 megabecquerels (MBq)<br>= 0.01 grey (Gy)<br>= 0.01 sievert (Sv) |
| 1 roentgen (R) | = 0.000258 coulomb/kilogram |
| 1 megabecquerel (MBq) | = 0.027 millicuries (mCi) |
| 1 grey | = 100 rad |
| 1 sievert (Sv) | = 100 rem |
| 1 coulomb/kilogram (C/kg) | = 3,880 roentgens |

*table contd....*

*table contd....*

**CONVERSION FACTORS**

| *To convert from* | *To* | *Multiply by* |
|---|---|---|
| Curies (Ci) | becquerels (Bq) | 3.7 x $10^{10}$ |
| millicuries (mCi) | megabecquerels (MBq) | 37 |
| microcuries (u*Ci)* | megabecquerels (MBq) | 0.037 |
| millirads (mrad) | milligreys (mGy) | 0.01 |
| millirems (mrem) | microsieverts (*u*Sv) | 10 |
| milliroentgens (mR) | microcoulombs/kg (*uC/kg)* | 0.258 |
| becquerels (Bq) | curies (Ci) | 2.7 x $10^{-11}$ |
| megabecquerels (MBq) | millicuries (mCi) | 0.027 |
| megabecquerels (MBq) | microcuries (*u*Ci) | 27 |
| milligreys (mGy) | millirads (mrad) | 100 |
| microsieverts (*u*Sv) | millirems (mrem) | 0.1 |
| microcoulombs/kg (*u*C/kg) | milliroentgens (mR) | 3.88 |

[a] See Glossary

figures were 163,000 Bq/kg of cesium-137 and 1,170,000 Bq/kg of iodine-131. On March 30th, iodine-131 concentrations in seawater 300 m south of a key discharge outlet had reached 180,000 Bq/mL, 3,355 times the legal limit in Japan. On April 3rd, mushrooms contained radioactive substances higher than the legal limit (Wikipedia 2011a).

Three weeks after the Fukushima accident, a clear trace reaching out 30 to 40 km northwest of the plant marked a zone of dose rate above 123 microsieverts per hour, a level at which immediate evacuation is often advised (Smith 2011). Cesium-137 deposition in this area reached a maximum of 3.7 megabecquerels per square meter (MBq/$m^2$; wherein 1 MBq = 1 x $10^6$ Bq = 27 microcuries) and this is consistent with Japanese soil data from Iitate village 40 km northwest of the plant. At that time, little information on cesium-137 contamination was available within 20 km of the plant (the distance of the evacuation zone), this could be of the order of megabecquerels per square meter if the isotopic composition of deposits near the plant is similar to that in the area farther to the northwest. If large areas are contaminated with 0.5 MBq/$m^2$ or more, evacuation could be for the long-term. Author concludes that food restrictions could hold for decades, particularly for wild foodstuffs such as mushrooms, berries and freshwater fish (Smith 2011).

In April, the U.S. Department of Energy published projections of radiation risks over the next year for people living in the Fukushima vicinity. Potential exposure could exceed 20 mSv/year (2 rems/year) in some areas up to 50 km from the plant (Wikipedia 2011c). This is the level at which relocation would be considered in the USA and it is a level projected to cause one extra cancer case in 500 young adults. As of September 11th there were no deaths or serious injuries due to radiation exposures although 6 workers had exceeded lifetime legal limits for radiation and more than 300 had received significant radiation doses (Wikipedia 2011c). On April 12, 2011, one month after a 9.0 magnitude earthquake and accompanying tsunami disabled four reactors at the Fukushima Daiichi Nuclear Power Station, the Nuclear and Industrial Safety Agency (NISA), Japan's regulatory agency, announced that the releases of radioactivity into the atmosphere qualified it as a "major accident" or a Level 7 emergency, the highest level on the International Radiological Event Scale (Hippel 2011). In June, 3 months after the incident, radioactivity releases into the atmosphere from the Fukushima accident were compared to the 1986 Chernobyl accident (Eisler 1995, 2003), the only other Level 7 accident in history. Atmospheric releases of radioactivity from Fukushima was about one tenth that of Chernobyl and the area contaminated with cesium-137 at the same levels that caused evacuation in Chernobyl was also about one tenth as large (Hippel 2011). The estimated number of resulting cancer deaths in the Fukushima area from contamination due to more than 1 curie/$km^2$ is likely to scale accordingly. This is not unreasonable since the Chernobyl releases went directly into the atmosphere while at Fukushima much of the radioactivity that was released from the reactors was captured in the water inside the reactor buildings or discharged into the ocean. Because contaminated milk was interdicted in Japan, the number of cancer cases will probably be less than 1% of similar cases in Chernobyl. Actual releases and cancer cases are as yet unknown, nor are the long-term psychological effects (Hippel 2011). The heavily-damaged and contaminated complex will probably be closed once the crisis is over (Wikipedia 2011; 2011c).

Radioisotopes of iodine, tellurium, barium, lanthanum, arsenic, chlorine, xenon, strontium, cesium, technetium, sulfur, uranium and plutonium were among those reported released to the environment; but the information is fragmented and heavily censored. Over time, the true picture will emerge of damage effects and mitigation strategies.

Radiation levels in Fukushima were lower than predicted (Whyte 2011) based on a study conducted between March 15th and June 20th of 2011 in which radiation levels of 5,000 residents of Fukushima and environs were measured and only 10 had elevated radiation levels that were still below the threshold associated with gastrointestinal tract damage (Figure 5.1). Author concludes that not much radioactive

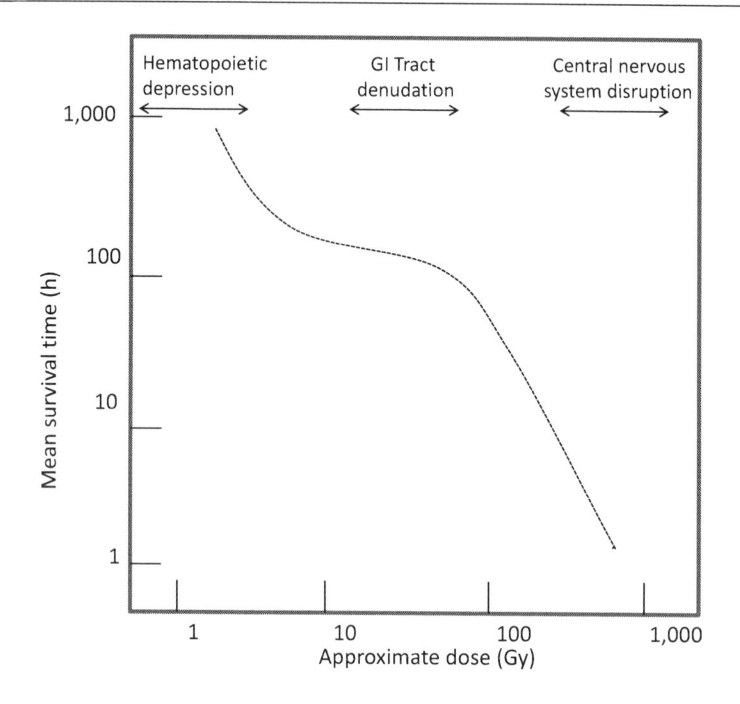


**Figure 5.1** Survival time and associated mode of death in humans after acute whole body doses of gamma radiation (modified from Hobbs and McClellan 1986 and UNSCEAR 1988).

material escaped from Fukushima when compared to Chernobyl (Whyte 2011). Others disagree and conclude that the Fukushima nuclear plant released far more radiation than government estimates (Brumfiel 2011), owing to under-reporting of spent nuclear fuel releases of cesium-137 which could have been prevented by prompt action (Brumfiel 2011).

## 5.2 Radiological Criteria

The current radiation regulatory level in Japan for occupational health is set at 250 mSv yearly (Table 5.2). Iodine-131 and cesium-137 levels in spinach, seawater and drinking water are also shown (Table 5.2). By comparison, radiological criteria for the protection of human health worldwide are briefly summarized for cancer risk, birth defects, diet, students, pregnant women and occupational workers (Table 5.3).

For all radionuclides mentioned herein, the nuclide, mass number and half-life are listed in Table 5.4.

Provisional regulation values for radioactivity in food and drink were set on March 17th (Hamada and Ogino 2011). For radiocesiums,

**Table 5.2 Current radiation regulatory levels in Japan (from Wikipedia 2011a).**

| *Sample* | *Level* |
|---|---|
| Occupational health | 250 mSv yearly |
| Spinach | 2 Bq/g of iodine -131 |
| Spinach | 0.5 Bq/g cesium-137 |
| Seawater at discharge | 0.04 Bq/mL iodine-131 |
| Seawater at discharge | 0.09 Bq/mL cesium-137 |
| Drinking water | 0.1 Bq/g iodine-131 |
| Drinking water | 0.2 Bq/g cesium-137 |

**Table 5.3. Selected radiological criteria for the protection of human health.**

| *Criterion and other variables* | *Concentration or dose* | *Reference*[a] |
|---|---|---|
| *Cancer risk, projected increase in cancers (USA)* | | |
| 0.04% | 0.11 mSv whole body/year | 1 |
| 0.18% | 0.51 mSv (0.05 rem) whole body/year | 1 |
| 0.92% | 3.3 mSv whole body/year | 1 |
| 4.4% | 20.1 mSv (2.0 rem whole body/year | 1 |
| *Birth defects, projected increase (USA)* | | |
| 0.01% | 0.69 mSv whole body/year | 1 |
| 0.07% | 3.3 mSv whole body over 30 years | 1 |
| *Diet* | | |
| Adults, Italy | 600 Bq cesium 134+137 kg FW total diet | 2 |
| Children, Italy | 370 Bq/kg FW total diet | 2 |
| Fish, Great Lakes (USA), muscle | dose of <20 mSv | 3 |
| Fish, Sweden | <1,500 Bq cesium-137/kg FW | 4 |
| Milk, Italy | 370 Bq cesium 134+137/L | 2 |
| Milk, Sweden | 300 Bq cesium-137/L | 5 |
| Meat and fish, Sweden | 1,500 Bq cesium-137/kg FW | 5 |
| *General Public. Maximum permissible dose (mSv/whole body/year)* | | |
| Students | <1 mSv | 6 |
| Pregnant women | <5 mSv during the first 2 months of pregnancy | 7 |
| All others | <5mSv | 3, 6, 7 |
| *General Public Annual effective dose*[b] *Cesium-137; total intake* | | |
| Sweden | <50,000 Bq/year or <1 mSv/year | 8 |

*table contd....*

*table contd....*

| | | |
|---|---|---|
| North America | <300,666 Bq/year | 9 |
| United Kingdom | <400,000 Bq/year equivalent to <5 mSv or 0.5 rem/year | 10 |
| *Occupational workers (Annual limit of intake[b]; inhalation vs. oral)* | | |
| Americium-241 | 200 Bq vs. 50,000 Bq | 11 |
| Cesium-137 | 6 million Bq vs. 4 million Bq | 11 |
| Cobalt-60 | 1 million Bq vs. 7 million Bq | 11 |
| Iodine-125 | 2 million Bq vs. 1 million Bq | 11 |
| Iodine-129 | 300,000 Bq vs. 200,000 Bq | 11 |
| Iodine-131 | 2 million Bq vs. 1 million Bq | 11 |
| Plutonium-239 | 200 Bq vs. 200,000 Bq | 11 |
| Radium-226 | 20,000 Bq vs. 70,000 Bq | 11 |
| Strontium-90 | 100,000 Bq vs. 1.0 million Bq | 11 |
| Uranium-235 | 2,000 Bq vs. 500,000 Bq | 11 |
| *Total annual intake from all sources, Canada* | | |
| Lead-210 | <4,000 Bq | 12, 13 |
| Polonium-210 | <10,000 Bq | 12, 13 |
| Thorium-228 | <50,000 Bq | 12, 13 |
| Thorium- 230 | <40,000 Bq | 12, 13 |
| Thorium-232 | <7,000 Bq | 12, 13 |
| Effective dose [b] | | |
| Average annual | 20 mSv, not to exceed 50 mSv | 14 |
| Five year maximum | <100 mSv | 14 |
| *Maximum permissible dose* | | |
| Whole body | 50 mSv per year | 6, 7, 15 |
| Pregnant women | 5 mSv in gestation period | 6 |
| *Soil* | | |
| Radium-226: maximum | <185 Bq over background in top 15 cm; <555 Bq in soils at depth>25 cm | 16 |
| Total gamma; maximum | 200 mSv over background | 16 |

[a] 1, Bair et al.1979; 2, Battiston et al. 1991; 3, Joshi 1991; 4, Hakanson and Andersson 1992; 5, Johanson et al. 1989; 6, Hobbs and McClellan 1986; 7, ICRP 1977; 8, Johanson 1990; 9, Allaye-Chan et al. 1990; 10, Lowe and Horrill 1991; 11, Kiefer 1990; 12, Clulow et al. 1991; 13, Clulow et al 1992; 14, ICRP 1991b; 15; ICRP 1991a; 16, CFR 1991.

[b] The Annual Limit of Intake (ALI) for any radionuclide is obtained by dividing the annual effective dose limit (20 mSv) by the committed effective dose (E) resulting from the intake of 1 Bq of that radionuclide. ALI data for individual radionuclides are given in ICRP (1991b).

**Table 5.4. Half-life and atomic number of selected radionuclides (modified from Eisler 2000).**

| *Nuclide* | *Atomic Number* | *Half-Life* |
|---|---|---|
| Beryllium-7 | 4 | 53.3 days |
| Sulfur-35 | 16 | 87 days |
| Chlorine-35 | 17 | stable |
| Potassium-40 | 19 | 1.25 billion years |
| Manganese-54 | 25 | 312 days |
| Iron-55 | 26 | 2.7 years |
| Iron-59 | 26 | 45 days |
| Cobalt-57 | 27 | 271 days |
| Cobalt-58 | 27 | 71 days |
| Cobalt 60 | 27 | 5.3 years |
| Zinc-65 | 30 | 244 days |
| Strontium-89 | 38 | 50.5 days |
| Strontium-90 | 38 | 29.0 years |
| Krypton-95 | 36 | 114 milliseconds |
| Zirconium-95 | 40 | 65 days |
| Niobium-95 | 41 | 35 days |
| Molybdenum-99 | 42 | 66 hours |
| Technetium-99m | 43 | 6 hours |
| Ruthenium-103 | 44 | 40 days |
| Ruthenium-105 | 44 | 4.4 hours |
| Ruthenium-106 | 44 | 373 days |
| Silver-110m | 47 | 250 days |
| Antimony-125 | 51 | 2.7 years |
| Tellurium-129m | 52 | 33 days |
| Tellurium-129 | 52 | 69.5 minutes |
| Tellurium-132 | 52 | 78.2 hours |
| Iodine-125 | 53 | 60 days |
| Iodine-131 | 53 | 8 days |
| Iodine-132 | 53 | 2.33 hours |
| Iodine-134 | 53 | 52,5 minutes |
| Iodine-135 | 53 | 6.6 hours |

*table contd....*

*table contd....*

| | | |
|---|---|---|
| Xenon-133 | 54 | 5.3 days |
| Cesium-134 | 55 | 2.06 years |
| Cesium-136 | 55 | 13.04 days |
| Cesium-137 | 55 | 30.2 years |
| Barium-140 | 56 | 12.8 days |
| Lanthanum-140 | 57 | 40 hours |
| Cerium-141 | 58 | 33 days |
| Cerium-144 | 58 | 284 days |
| Lead-210 | 82 | 22.3 years |
| Radium-226 | 88 | 1,620 years |
| Thorium-228 | 90 | 1.91 years |
| Thorium-230 | 90 | 75,400 years |
| Thorium-232 | 90 | 14 billion years |
| Uranium-234 | 92 | 125,000 years |
| Uranium-235 | 92 | 710 million years |
| Uranium-238 | 92 | 1.47 billion years |
| Neptunium-239 | 93 | 2.35 days |
| Plutonium-238 | 94 | 87.7 years |
| Plutonium-239 | 94 | 24,110 years |
| Plutonium-240 | 94 | 6,537 years |
| Plutonium-241 | 94 | 14.4 years |
| Americium-241 | 95 | 458 years |
| Curium-242 | 96 | 463 days |
| Curium-243 | 96 | 28.5 years |
| Curium-244 | 96 | 18.1 years |

uranium, plutonium and transuranic alpha emitters, values in food should not exceed a committed effective dose of 5 mSv per year. Values for radioiodine in water and milk and some vegetables, however, should be less than a committed equivalent dose to the thyroid of 50 mSv per year. Index values were calculated as radioactive concentrations of indicator radionuclides (iodine-131, cesium 134 and cesium-137) by postulating the relative radioactive concentration of coexisting radionuclides (e.g., iodine-132, iodine-133, iodine-134, iodine-135 and tellurium-132). Provisional regulation levels were exceeded in tap water, raw milk and some vegetables; accordingly, restrictions on distribution and consumption began on March 21st. Fish contaminated with radioiodines were detected on March 21st; provisional regulation values for radiodines in seafood, adapted from that in vegetables, were set on April 5th. Overall, restrictions started within 25 days after provisional radiocontamination values were exceeded in food and drink (Hamada and Ogino 2011).

Survival time of humans after acute exposure to a whole body dose of 100,000 mSv gamma radiation is about 100 hours, with hematopoietic depression the most sensitive indicator of radiation stress, followed by GI tract denudation, then central nervous system disruption (Figure 5.1). Humans were more sensitive than monkeys, mice, goats and pigs (Eisler 2000).

## 5.3 Atmospheric Releases

Major atmospheric releases of iodine-131 (more than $10^{15}$ Bq per hour) were estimated during the afternoon of March 12th after the hydrogen explosion of Unit 1 and late at night on March 14th (Katata et al., 2012). The release rate in other periods from March 12th through March 15th was on the order of $10^{13}$ Bq per hour. The high concentration plumes discharged during these periods (containing iodine-131 as well as cesium-137) flowed in northwesterly and south-southwesterly directions, causing a large amount of dry deposition on the ground surface (Katata et al., 2012).

Following a massive release of radioactive material on March 14th from the Fukushima reactors, the vertical atmospheric DC electric field at ground level, or potential gradient (PG, or voltage per unit length), dropped by one order of magnitude at Kakioka, 150 km southwest of the facility (Takeda et al., 2011). The PG stayed depressed for several days and is similar to a PG drop associated with rain-induced fallout after nuclear tests and the Chernobyl disaster. Arrival of radioactive dust by low altitude wind without rain also caused a PG drop. Authors recommend that all nuclear power plants have a network of PG observation sites surrounding the plant (Takeda et al., 2011).

Hydrogen and vapor blasts that occurred until March 15th outside the Fukushima reactors led to the emission of radioactive materials into the atmosphere (Takemura et al., 2011). Long-range transport to Europe and the United States was simulated by an atmospheric global transport model. Radioactive particles could cross the Pacific Ocean within 3 to 4 days, with almost all radioactivity lost which is consistent with the level detected in California on March 18th. The model also reproduces the subsequent trans-Atlantic transport of these particles by a poleward-deflected jet stream, first toward Iceland and then southward to continental Europe as actually observed (Takemura et al., 2011) .

Radioactive sulfur-35 measured in sulfate aerosols and sulfur dioxide gas at a coastal site in La Jolla, California, was 365 times above expected natural concentrations and suggested that nearly 4 x $10^{11}$ neutrons per $m^2$ leaked from the Fukushima nuclear power plant before March 20th (Priyadarshi et al., 2011). A significantly higher radiosulfate activity was measured on March 28th and is in accord with neutrons escaping the reactor core and being absorbed by the coolant seawater chlorine-35 to produce sulfur-35, which subsequently oxidizes to $^{35}SO_2$ and $^{35}SO_4^{2-}$ which were then transported to southern California by prevailing winds (Strain 2011; Priyadarshi et al., 2011). Priyadarshi et al. (2011) conclude that approximately 0.7% of the total radioactive sulfate present at the marine boundary layer at Fukushima reached southern California as a result of the trans-Pacific transport (Priyadarshi et al., 2011). Strub et al. (2011) acknowledged that the detected sulfur-35 originated from Fukushima, but expressed doubts about the proposed creation and release path and presented several alternative sources.

The spread of nuclear radiation from the power plant was predicted with a short-term climate forecasting model (Qiao et al., 2011). Results show that radiation released relatively close to the surface would cover North America 10 days after the initial leakage at markedly reduced concentrations. Europe would be blanketed in 15 days and much of the northern hemisphere in 30 days. If the initial leakage is assumed to occur at 5,000 m aboveground, Europe would be blanketed in 10 days and much of the northern hemisphere in 15 days. Radiation released at 10,000 m aboveground, would affect much of China after 10 days (Qiao et al., 2011). Atmospheric transport and deposition of iodine-131 and cesium-137 were simulated using a chemical transport model (Morino et al., 2011). The model approximated the observed temporal and spatial variation rates over 15 Japanese prefectures, including Tokyo, 60 to 400 km from Fukushima. Discrepancies between simulated and observed rates were attributed to uncertainties in the simulation of emission, transport and deposition processes in the model. Authors concluded that 13% of the iodine-131 and 22% of the cesium-137 were deposited over land in Japan and the rest deposited over the ocean or more than 700 km distant (Morino et al., 2011).

As of June 6, 2011, an estimated 700,000 trillion Bq of radiation had escaped to the atmosphere from Fukushima; the highest radiation level of 10,000 mSv/h was recorded on August 1st in ventilation shafts between reactors 1 and 2 (Table 5.5). On August 24, 2011, the Nuclear Safety Commission of Japan stated that the total amount of radioactive

**Table 5.5. Radiation releases to the atmosphere from Fukushima over time. Modified from Osnos (2011) and Wikipedia (2011).**

| *Date* | *Measured radiation levels in air surrounding Fukushima* |
|---|---|
| March 15,2011 | 400 millisieverts (mSv) per hour (h) from radioisotopes of strontium, technetium, and possibly uranium |
| March 16 | 1,000 mSv/h with no radiation detected 80 km distant |
| March 17 | Total release since March 11 estimated at 770 trillion becquerels (Bq), mostly iodine radioisotopes |
| March 18 | 0.15 mSv/h measured 30 km northwest of Fukushima |
| March 24 | 0.2 mSv/h |
| March 27–31 | Total released is about 800 million Bq |
| April 3 | 1, 000 mSv/h. Daiichi plants may continue to release dangerous levels of radiation into the atmosphere for several months |
| April 7 | >100,000 mSv/h where 10,000 mSv/h is fatal in less than 30 minutes |
| April 8 | 0.65 mSv/h |
| April 12 | 84 mSv/h Total released since March 12$^{th}$ -190,000 trillion Bq, mostly radioiodine |
| April 18 | 57 mSv/h inside unit 3 |
| April 27 | 1,120 mSv/h; reading by robot |
| May 15 | 2,000 mSv/h; robot reading as 8 min is maximum at this level for human worker safety |
| June 4 | 4,000 mSv/h in reactor 1 building |
| June 6 | 770,000 trillion Bq is new estimate of escaped radiation |
| June 21 | 430 mSv/h at reactor 2 |
| August 1 | 10,000 mSv/h in ventilation shafts between reactors 1 and 2 and is the highest level recorded since the initial March explosion |
| August 2, 2011 | 5,000 mSv/h on second floor of Turbine building; considered lethal for brief human exposure |

[a] Original Japanese occupational standard was less than 100 mSv over a 5-year period. Since Fukushima this value was raised to 250 mSv. Of nearly 4,000 workers who have passed through the plant since the meltdowns, only 103 have received more than 100 mSv of radiation. The most serious cases are 2 workers who were exposed to >500 mSv in the first weeks of the crisis.

materials released into the atmosphere during the Fukushima accident between March 11th and April 5th included 1.3 x $10^{17}$ Bq of iodine-131 and 1.2 x $10^{16}$ Bq of cesium-137 (Wikipedia 2011c). New calculations of total cesium-release yielded 35.8 x $10^{15}$ Bq or about 42 percent of the estimated Chernobyl emission (Stohl et al., 2011). In addition, a total estimated release of 16.7 x $10^{18}$ Bq of xenon-133 (uncertainty range of 13.4 to 20 x $10^{18}$ Bq) which is the largest radioactive noble gas release in history not associated with nuclear bomb testing (Stohl et al., 2011). There is strong evidence that the first major release of radioxenon started immediately after the earthquake. The entire noble gas inventory of reactor units 1–3 was released into the atmosphere between March 11th and March 15th.

## 5.4 Aquatic Releases

The radioactive water treatment facility was shut down after a filter exceeded the radioactivity limit (Wikipedia 2011). The separation unit—which removes cesium from the water—exceeded its limit after only 5 hours of operation instead of the usual month . This was attributed to radioactive oil and sludge in the water which contained more radioactivity than expected (Wikipedia 2011). By March 25th seawater near the plant contained 50,000 Bq/L of iodine-131; a value 1,250 times over background (Table 5.6). Samples of seawater collected near the Fukushima reactors extending as far as 30 km offshore contained radiation levels by late March that were tens of thousands of times higher than before the accident (Schiermeier 2011). But the total amount of radioactivity that entered the ocean by late March is unknown and discharges—both accidental and deliberate—are continuing. About 50 different radioisotopes contributed to an overall concentration of about 10,000 Bq/L in seawater within 300 m of Fukushima, mainly isotopes of radiocesium and radioiodine. Researchers suggest that sediments in the region in early April may contain as much as 10 million Bq/kg (Schiermeier 2011).

On March 25th, stagnant water in the basement floor of the turbine building of unit 1 was analyzed and found to be heavily contaminated with radioisotopes of chlorine, arsenic, yttrium, iodine, cesium-134, cesium-136, cesium-137 and lanthanum (Wikipedia 2011a). By April 5th, seawater radioactivity near the plant was 7.5 million times the legal limit, with most radioactivity attributed to iodine-131 (Table 5.6). On April 6th, 12,700 tons of contaminated wastewater was deliberately discharged into the Pacific Ocean despite protests from commercial

**Table 5.6. Radiation releases to the aquatic environment from Fukushima over time. Modified from Wikipedia (2011).**

| *Date* | *Measured radiation* |
|---|---|
| March 23, 2011 | Tokyo drinking water. Elevated. Twice safe level for baby formula |
| March 24 | Seawater near reactors. 3.9 billion Bq/L |
| March 25 | Seawater near plant. Iodine-131 level is 50,000 Bq/L or 1,250 times background |
| March 27 | Coolant water within Unit 2 reactor; 750-1,000 mSv/h, possibly iodine-134 |
| March 30 | Seawater near plant; radiation levels 4,385 times higher than regulatory limit mainly from radio isotopes of iodine |
| April 1 | Groundwater. Radioiodine isotopes 10,000 times higher than recommended level |
| April 2 | Contaminated reactor water leaking directly onto sea from cracks in the wall; >1,000 mSv/h; radioiodine levels 7.5 million times higher than regulatory level |
| April 3 | Seawater; 1,000 mSv/h; radiocesium levels 1.1 trillion times legal limit |
| April 4 | Seawater near plant. Tokyo Electric Power Company proposes to dump 11,500 tons of low level radioactive wastewater (total radioactivity is 4 times the legal limit) from storage tanks directly into the Pacific Ocean. Radiocesium in wastewater is 1.1 million times legal limit. |
| April 5 | Seawater radioactivity near plant is 7.5 million times legal limit, Radioactivity attributed to iodine-131 from leaky pipe in Unit 2. |
| April 6 | 12,700 tons of contaminated wastewater discharged into the sea despite protests from fishermen of Russia, South Korea, and Japan |
| April 9 | Radiation levels in seawater exceed legal limit by several thousand times. With most radioactivity contributed by by various radioiodine species and from cesium-134 and cesium-137 |
| May 8 | Maximum levels of radiocesium and radioiodine recorded |
| May 10-22 | 250 tons of contaminated water from reactor 3 leaked directly into the Pacific Ocean over a period of 41 hours |
| June 29 | Seawater sample near water intake of reactor 1 contained 720 Bq /L of tellurium-129 (2.4 times safe level. Half-life of this isotope is 34 days |
| July 4, 2011 | Tokyo tapwater contains elevated levels of cesium-137 (half-life of 30 years) for first time since April |

fishing interests from Russia, South Korea and Japan (Wikipedia 2011; Table 5.5). Other examples of radioactive waste disposal at sea by the United States, the United Kingdom and Europe are shown in Table 5.7.

On April 21st TEPCO estimated that 520 tons of radioactive water had leaked into the sea before leaks were plugged, releasing 4,700

**Table 5.7. Some examples of radioactive waste disposal at sea.**

| *Disposer and other variables* | *Quantity, in trillions of becquerels (TBq)* | *Reference*[a] |
|---|---|---|
| **UNITED STATES** | | |
| Atlantic Ocean, 1951-60 vs. 1961-67 | 2,939 vs. 2 | 1 |
| Pacific Ocean, 1951-60 vs. 1961-67 | 527 vs. 16 | 1 |
| **UNITED KINGDOM** | | |
| 1951–67, alpha vs. beta | 123 vs. 1,631 | 1 |
| Sellafield, alpha (primarily plutonium and americium | | |
| 1968–70 | 50 vs. 61 | 2 |
| 1971 vs. 1972 | 90 vs. 143 | 2 |
| 1973 vs. 1974 | 181 vs. 17 | 2 |
| Sellafield reprocessing plant[b] | | |
| 1980 vs. 1981 | 5,145 vs. 4,451 | 3 |
| 1982 vs. 1983 | 4,005 **vs.** 3,112 | 3 |
| 1984 vs. 1985 | 1,835 vs. 646 | 3 |
| **EUROPE** | | |
| Germany. Belgium, Netherlands, France; 1961; alpha vs beta plus gamma | 6 vs. 220 | 1 |
| France, Cap de la Hague processing plant[c] | | |
| 1980 vs. 1981 | 503 vs. 455 | 3 |
| 1982 vs. 1983 | 693 vs. 683 | 3 |
| 1984 vs. 1985 | 670 vs. 674 | 3 |

[a] 1= Joseph et al. 1971; 2= Hetherington et al. 1976; 3=UNSCEAR 1988.
[b] Effluent composition primarily cesium-137 and plutonium-241.
[c] Effluent composition primarily ruthenium-106 and antimony-125.

trillion Bq of total water release, including 2,800 trillion Bq of iodine-131, 940 trillion Bq of cesium-134, 940 trillion Bq of cesium-137, as well as minor amounts of technetium-99m, tellurium-129m, barium-140 and lanthanum-140 (Wikipedia 2011a). Peak ocean discharges of radionuclides were recorded in early April followed by a 1,000-fold decrease in May; however concentrations through the end of July remained higher than expected, suggesting continued releases from the reactors or other contaminated sources such as groundwater or coastal sediments (Buesseler et al., 2011). By July, levels of cesium-137 were still more than 10,000 times higher than levels measured in 2010 in Japanese coastal waters. Although some radionuclides were significantly elevated, dose calculations suggest minimal impact on marine biota or humans at this time. Further studies appear needed on biological uptake and effects of consuming contaminated seafood (Buesseler et al., 2011). On September 8, 2011, Japanese scientists from the Japan Atomic Energy Agency, Kyoto University and other institutes stated that the total amount of radioactive material released into the ocean between late March through April was 15,000 trillion Bq for iodine-131 plus cesium-137; this was more than triple the figure of 4,720 T Bq estimated by TEPCO (Wikipedia 2011c). The spread of Fukushima-derived radiation through the ocean was predicted with an ocean circulation model (Qiao et al., 2011). The circulation model indicates that the nuclear material would be slowly transported northeast of Fukushima reaching 150o E in 50 days and would be confined to a comparatively narrow band (Qiao et al., 2011).

From March 26th to the end of May, the total amount of cesium-137directly discharged into the surrounding ocean was estimated at 3.5 PBq = $3.5 \times 10^{15}$ Bq (Tsumune et al., 2011). Radioactive water will continue to flow from the cores into basements and trenches, eventually ending up in the ocean (Brumfiel and Cyranoski 2011). TEPCO eventually plugged the leaks using a compound known as liquid glass (Mufson 2012).

## LITERATURE CITED


Allaye-Chan, A.C., R.G. White, D.F. Holleman and D.E. Russell. 1990. Seasonal concentrations of cesium-237 in rumen content, skeletal muscles and feces of caribou from the Porcupine herd; lichen ingestion rates rates and implications for human consumption. *Rangifer,* Special Issue 3, pp. 17–24.

Bair, W.J., J.W. Healy and B.W. Wachholz. 1979. The Enewetak Atoll today. U.S. Department of Energy, Washington, D.C. 25 p.

Battiston, G.A., S. Degetto, R. Gerbasi, G. Sbrignadello, R. Parigi-Bini, G. Xiccato and M. Cinetto. 1991. Transfer of Chernobyl fallout radionuclides feed to growing rabbits: cesium-137 balance. *The Science of the Total Environment.* **105:**1–12.

Brumfiel, G. 2011. Fallout forensics hike radiation toll, *Nature.* **478:**435–436. Doi:10.1038/478435a. .

Brumfiel, G. and D. Cyranoski. 2011. Fukushima deep in hot water. *Nature.* **474:**135–136.

Buesseler, K., M. Aoyama and M. Fukasawa. 2011. Impacts of the Fukushima nuclear power plants on marine radioactivity, *Environmental Science & Technology.* **45(23):**9931–9935.

CFR (United States Code of Federal Regulations). 1991. *Subchapter FB Radiation Protection Programs*. **40:**6–23.

Clulow, F.V., M.A. Mirka, N.K. Dave and T.P. Lim. 1991. $^{226}$Ra and other radionuclides in water, vegetation and tissues of beavers (*Castor canadensis*) from a watershed containing U tailings near Elliot Lake, Canada, *Environmental Pollution*, **69:**277–310.

Clulow, F.V., T.P. Lim, N.K. Dave and R. Avadhanula. 1992. Radium-226 levels and concentration ratios between water, vegetation and tissues of ruffed grouse (*Bonasa umbellus*) from a watershed with uranium tailings near Elliot Lake, Canada, *Environmental Pollution*. **77:**39–50.

Eisler, R. 1995. Ecological and toxicological aspects of the partial meltdown of the Chernobyl nuclear plant reactor. In: D.J. Hoffman, B.A. Rattner, G.A. Burton, Jr. and A. Cairns Jr. (eds.) *Handbook f Ecotoxicology*, Lewis Publishers, Boca Raton, Florida, pp. 459–564.

Eisler, R. 2000. Radiation. In Handbook of Chemical Risk Assessment, Volume 3. Lewis Publishers, Boca Raton, Florida, pp. 1707–1828.

Eisler, R. 2003. The Chernobyl nuclear power plant reactor accident: ecotoxicological update. In: D.J. Hoffman, B.A. Rattner, G.A. Burton, Jr. and A. Cairns Jr. (eds.). Handbook of Ecotoxicology, Second Edition, Lewis Publishers, Boca Raton, Florida, pp. 703–736.

Hakanson, L. and T. Andersson. 1992. Remedial measures against radioactive caesium in Swedish lake fish after Chernobyl, *Aquatic Science*. **54:**141–164.

Hamada, N. and H. Ogino. 2011. Food safety regulations: What we learned from the Fukushima nuclear accident, *Journal of Environmental Radioactivity*, doi:10.1016/j.jenvrad.2011.08.008.

Hetherington. J.A., D.F. Jefferies. N.T. Mitchell, R.J. Pentreath and D.S. Woodhead. 1976. Environmental and public health consequences of the controlled disposal of transuranic elements to the marine environment. pp. 139–154 in Transuranium nuclides in the environment, IAE-SM-1999/11. International Atomic Energy Agency, Vienna.

Hippel, F.N.V. 2011. The radiological and psychological consequences of the Fukushima Daiichi accident, *Bulletin of the Atomic Scientists*. **67(5):**27–38.

Hobbs, C.H. and R.O. McClellan. 1986. Toxic effects of radiation and radioactive materials. pp. 669–705 in C.D. Klaassen, M.O. Amdur and J. Doull (eds.), Casarett and Doull's toxicology. Third edition. Macmillan, New York.

ICRP (International Commission on Radiological Protection). 1977. Recommendations of the international commission on radiological protection. *ICRP publication 26, Annals of the ICRP.* **1(3):**1–53.

ICRP (International Commission on Radiological Protection). 1991a. 1990 recommendations of the international commission on radiological protection. *ICRP publication 60, Annals of the ICRP.* **211-(3):** 1–201.

ICRP (International Commission on Radiological Protection). 1991b.

Annual limits on intake of radionuclides by workers based on the 1990 recommendations. *ICRP publication 61, Annals of the ICRP.* **21(4):**1–41.

Johanson, K.J. 1990. The consequences in Sweden of the Chernobyl accident. *Rangifer, Special Issue.* **3:**9–10.

Johanson, K.J., G. Karlen and J. Bertilsson. 1989. The transfer of radiocesium from pasture to milk. *The Science of the Total Environment.* **85:**73–80.

Joseph. A.B., P.F. Gustafson, I.R, Russell, E.A. Schuert, H.L. Volchok and A. Tamplin. 1971. Sources of radioactivity and their characteristics. pp. 6–41 in National Academy of Sciences. *Radioactivity in the marine environment.* National academy of Sciences, Panel on radioactivity in the marine environment, Washington, D.C.

Joshi, S.R. 1991. Radioactivity in the Great Lakes. *The Science of the Total Environment.* **100:**61–104.

Katata, G., N. Ota. H. Tarada. M. Chino and H. Nagai. 2012. Atmospheric discharge and dispersion of radionuclides during the Fukushima Daiichi nuclear power plant accident. Part I: source term estimation and local-scale atmospheric dispersion in early phase of the accident, *Journal of Environmental Radioactivity.* **109:**103–113.

Kiefer, J. 1990. Biological radiation effects. *Springer-Verlag, Berlin,* 444 p.

Landers, P. 2012. Bad data guided U.S. Fukushima call, *Wall Street Journal (newspaper),* February 22nd, A 3.

Lowe, V.P.W. and A.D. Horrill. 1991. Caesium concentration factors in wid herbivores and the fox (*Vulpes vulpes* L.), *Environmental Pollution.* **70:**93–107,

Morino, Y, T. Ohara and M. Nishizawa. 2011. Atmospheric behavior, deposition and budget of radioactive materials from the Fukushima Daiichi nuclear power plant in March 2011. *Geophysical Research Letters,* 38, LOOG11, doi:10.1029/2011GL048689.

Mufson, S. 2012. The NRC's Japan meltdown, *Washington Post (newspaper),* February 7th, A 21.

Osnos, E. 2011. Letter from Fukushima. The fallout. *The New Yorker,* October 17, pp. 46–61.

Priyadarshi, A., G. Dominguez and M.H. Thiemens. 2011. Evidence of neutron leakage at the Fukushima nuclear plant from measurements of $^{35}S$ in California, *Proceedings of the National Academy of Sciences of the United States.* **108:**14422–14425.

Qiao, F.L., G.S. Wang, W. Zhao, J.C. Zhao, D.J. Dai, Y.J. Song and Z.Y. Song. 2011. Predicting the spread of nuclear radiation from the damaged Fukushima nuclear power plant, *Chinese Science Bulletin.* **56:**1890–1896.

Schiermeier, Q. 2011. Radiation release will hit marine life, *Nature.* **472:**145–146

Smith, J. 2011. A long shadow over Fukushima, *Nature.* **472:**7, doi:10:1038/472007a.

Stohl, A., P. Seibert, G. Wotawa, D. Arnold, J.F. Burkhart, S. Eckhardt, C. Tapia, A. Vargas and T.J. Yasunari. 2011. Xenon-133 and caesium-137 releases into the atmosphere from the Fukushima Dai-ichi nuclear power plant:

determination of the source term, atmospheric dispersion and deposition, *Atmospheric Chemistry and Physics Discussions.* **11:**28319–28394, doi:10.5194/acpd-11-28319-2011,2011

Strub, E., B. Gmal, V. Hannstein, G. Pretzsch and E. Schrodt. 2011. Creation path of $^{35}S$ from Fukushima not so obvious, *Proceedings of the National Academy of Science.* **108(51):**E1388.

Takeda, M., M. Yamauchi, M. Makino and T. Owada. 2011. Initial effect of the Fukushima accident on atmospheric electricity, *Geophysical Research Letters,* **38:**L15811. DOI:10.1029/2011gl048511.

Takemura, T., H. Nakamura, M. Takigawa, H. Kondo, T. Satomura, T. Miyasaka and T.A. Nakajima. 2011. A numerical simulation of global transport of atmospheric particles emitted from the Fukushima Daiichi nuclear power plant, *Sola.* **7:**101–104.

UNSCEAR (United Nations Scientific Committee on the Effects of Atomic Radiation), 1988. *Sources, effects and risks of ionizing radiations,* United Nations, New York, 647 p.

Whyte, C. 2011. Radiation levels in Fukushima re lower than predicted, PloSOne, DOI:10.371/journal.pone.002761.

Wikipedia. 2011, Timeline of the Fukushima Daiichi nuclear disaster, 21 p.

Wikipedia. 2011a. Radiation effects from Fukushima I nuclear accidents. 26 p.

Wikipedia. 2011c. Fukushima Daiichi nuclear disaster, 54 p.

Chapter 6

# Radiation Monitoring

## 6.1 General

On April 22nd the Japanese Ministry of Education, Culture, Sports, Science and Technology (MEXT) issued a press release on environmental monitoring with the objectives of obtaining an overview and providing data necessary to support the decision to establish evacuation routes (IAEA 2011). The plan included the following: data collection on radioactive material inside an appropriate area, including the area near the Fukushima Daiichi plant; continuing evaluations of changes in dose rates and accumulated radioactive materials in zones near the Fukushima plant; providing maps and other information to local residents on dose rates, distributions of radioactivity, estimated accumulated doses and levels of soil surface contamination. The enforced plan on environmental monitoring will be conducted in close cooperation between MEXT, the Japan Atomic Energy Agency, various universities, the Ministry of Defense, the local police and prefectural police, Fukushima prefecture, electrical utilities and others, including the U.S. Department of Energy and the International Atomic Energy Agency (IAEA 2011).

Since the Fukushima emergency, there has been great concern about human exposure to radionuclides, the evacuation and settling of those whose communities were affected, how and via what basis those decisions were made, concerns with crop safety from consumption to export, the issue of remediation, the extent and nature of areas affected (i.e., schools, playgrounds, roads, buildings, personal property), cost effectiveness and timeliness (Calabrese 2011).These factors raise significant questions and all are occurring amidst a swirl of governmental activities and multinational interactions, under the media spotlight and within an economic, political and social context (Calabrese 2011).

## 6.2 Atmosphere

On March 19th, the radioactive plume was advected over eastern Honshu Island where precipitation deposited a large fraction of cesium-137 on land surfaces (Stohl et al., 2011). The plume dispersed quickly over the entire northern hemisphere, first reaching North America on March 15th and Europe on March 22nd. It is estimated that 6.4 trillion Bq of cesium-137, or 19% of the total cesium fallout through April 20th, was deposited over Japanese land areas, while most of the rest fell over the North Pacific Ocean, Only 0.7 trillion Bq, or 2% of the total cesium-137 fallout, was deposited on land areas other than Japan (Stohl et al., 2011).

Aerosol monitoring at Tsukuba began on March 31st (Kanai 2011). Radionuclides such as molybdenum-99, technetium-99m, tellurium-129, tellurium-129m, tellurium-132, iodine-131 , iodine-132, cesium-134, cesium-136, cesium-137, barium-140, lanthanum-140, silver-110m and niobium-95 were detected and with the exception of cesium-137 and cesium-134, decreased to below the limit of detection by mid-June (Kanai 2011). The apparent atmospheric residence time of the Fukushima-derived cesium-137 at sites within 300 km of the reactor site is about 10 days (Hirose 2011).

Radioactive substances discharged into the atmosphere first reached the Chiba metropolitan area on March 15th (Amano et al., 2011). During the first two months, maximum daily concentrations of airborne radionuclides were as follows: 47.0 Bq/m$^3$ of iodine-131, 7.5 Bq/m$^3$ of cesium-137 and 6.1 Bq/m$^3$ of cesium-134. Observed daily deposit rate maxima were: $1.7 \times 10^4$ Bq/m$^2$ of iodine-131, $2.9 \times 10^3$ Bq/m$^2$ of cesium-137 and $2.9 \times 10^3$ Bq/m$^2$ of cesium-134 (Amano et al., 2011). Air samples at Fukushima were collected with a high volume aerosol sampler designed and developed at the Radiation Protection Bureau, Health Canada (Zhang et al., 2011). Iodine-131 and cesium-137 were detected in the air within 1 km of the plant on April 26th with maximum values of $11.8 \times 10^4$ Bq/cm$^3$ for total iodine-131 and $2.7 \times 10^4$ Bq/cm$^3$

for total cesium-137 (IAEA 2011). However, models of short- and long-term dispersion patterns of radionuclides in the atmosphere around the Fukushima nuclear power plant showed significant differences among plume directions in each month and season of the year (Leelossy et al., 2011)

Radioiodine-131 was first detected in aerosols from Korea on March 28th (Kim et al., 2011). By April 28th, maximum concentrations measured in aerosols were 3.12 mBq/m$^3$ for iodine-131 at Gunsan and 1.19 mBq/m$^3$ for cesium-134 and 1.25 mBq/m$^3$ at Busan (Kim et al., 2011).

Fission products started to arrive in the United States via atmospheric transport on March 15th, 2011 and peaked by March 23rd, 2011 (Biegalski et al., 2011). At that time, the maximum atmospheric concentration of iodine-131 was 300 Bq/m$^3$ in Melbourne, Florida; for xenon-133 it was 17 Bq/m$^3$ in Ashland, Kansas. While these levels are not health concerns, they were well above the detection capability of the radionuclide monitoring systems within the International Monitoring System of the Comprehensive Nuclear-Test-Ban Treaty (Biegalski et al., 2011). Fukushima-derived radioxenon, radiosulfur, radiotellurium, radioiodine and radiocesium isotopes were detected in the United States after the accident. On March 16th, 4-days after the Fukushima nuclear releases, xenon-133 was detected in the atmosphere over Seattle, Washington, more than 7,000 km distant (Bowyer et al., 2011). Maximum concentrations of xenon-233 were in excess of 40 Bq/m$^3$, being more than 40,000 times the average concentration of this isotope at this location (Boyer et al., 2011). On March 17th and 18th, 2011, radioactive fallout including isotopes of iodine-131, iodine-132, tellurium-132, cesium-134 and cesium-137 were detected in air filters at the University of Washington in Seattle (Leon et al., 2011; Wikipedia 2011a). Rainwater over San Francisco Bay contained trace amounts of Fukushima fallout products of iodine-131, iodine-132, tellurium-132, cesium 134 and cesium-137; the activity levels were low and posed no threat to the public (Norman et al., 2011). Radioxenon-133 was also detected in the air on March 20th, 2011 just off the west coast of Vancouver Island, Canada, in the range of 30 to 70 Bq/m$^3$ (Sinclair et al., 2011).

Radionuclides—including iodine and cesium isotopes—released into the atmosphere from the damaged Fukushima nuclear reactors were transported across the Pacific toward the North American continent and reached Europe despite dispersion and washout along the route of the contaminated air masses, with the first peak of activity level detected March 28th–March 30th (Masson et al., 2011). A rough estimate of the total iodine-131 inventory that passed over Europe in March was less than 1% of the released amount. Authors concluded that airborne activity levels were not a public health hazard in Europe (Masson et al., 2011). Iodine-131, cesium-134 and cesium-137 were recorded in air and rainwater from northern Greece between March 24th and April 9th 2011; the maximum concentration for iodine-131 in air was 497 $u$Bq/m$^3$ and for rainwater it was 0.7 Bq/L; maximum values for cesium radioisotopes during that same period in air was 145 $u$Bq/m$^3$ for cesium-137 and 126 $u$Bq/m$^3$ for cesium-134 (Manolopoulou et al., 2011). Radioiodine-131 arrived in Greece on March 24th (Potiriadis et al., 2011). In aerosols, the maximum iodine-131 concentration measured in Athens was 585 $u$Bq/m$^3$ and in Thessaloniki it was 408 $u$Bq/m$^3$; concentrations in gas were about 3.5 times higher than in aerosols. By April 29th, iodine-131 was below detection limits, as were cesium isotopes by May 16th (Potiriadis et al., 2011).

Traces of radioiodine and radiocesium isotopes were found in rainwater from Germany between March and May 2011 (Pittauerova et al., 2011). Elevated concentrations of several manmade radionuclides, including iodine-131, iodine-132, tellurium-132, cesium-134 and cesium-137, were detected in the atmosphere along the Iberian peninsula between March 28th and April 7th, 2011; analysis of back-trajectories of air masses demonstrated that the Fukushima nuclear accident was the source (Lozano et al., 2011). Iodine-131 and beryllium-7 isotopes discharged from Fukushima were detected in the atmosphere at Linz, Austria (Ringer et al., 2011)

Between March 14th and April 14th 2011, the following isotopes were detected by gamma-ray spectrometry in the air above-ground in Vilnius, Lithuania: tellurium-129m, tellurium-132 (in equilibrium with its daughter iodine-132), iodine-131, cesium-134, cesium-136 and cesium-137 (Gudelis et al., 2012). Cesium-137 increased from a background concentration of 1.6 $u$Bq/m3 to 893 $u$Bq/m$^3$. The maximum aerosol-associated iodine-131 activity concentration of 3,450 $u$Bq/m$^3$ was thousands of times lower than that measured in the same location in April-May 1986 as a consequence of the Chernobyl accident. The estimated gaseous fraction of iodine-131 comprised about 70 percent of the total iodine-131 activity (Gudelis et al., 2012).

## 6.3 Humans

### 6.3.1 *General*

Gamma dose rates were measured daily in all 47 Japanese prefectures. Values, in general, have decreased since March 20th. For Fukushima prefecture, mean gamma dose rates remained at 0.0018 mSv/hour. In Ibaraki prefecture gamma dose rates were slightly below 0.00012 mSv/hour; the other 45 prefectures had gamma dose rates below 0.00010 mSv/hour, falling within the range of local natural background levels. In Eastern Fukushima prefecture, for distances >30 km from the plant, a generally decreasing trend was documented, with values on April 26th ranging from 0.0001 to 0.0136 mSv/hour (IAEA 2011). Between March 15th and June 30th, radiation contamination levels of more than 5,000 residents forced to evacuate from a 20-km radius of the reactors had external contamination levels that were assessed as "no contamination level" (Monzen et al., 2011). Subsequently, the support group that measured external radiation in residents were themselves measured for internal radiation exposure using a whole body counter; results indicated undetectable levels in all staff members (Monzen et al., 2011).

Frustrated by a dearth of information on the fate of isotopes released from Fukushima, civic groups and individuals have monitored radiation on their own (Normile 2011). Collectively they have produced a worrisome picture of contamination throughout eastern Japan with some contaminated areas surprisingly far from the crippled reactors. It shows one wide belt of radiation reaching 225 km south of the reactors to Tokyo and another extending to the southwest, with localized areas high in radiation including northeast Tokyo and neighboring cities in Chiba Prefecture. Radiation in this zone is about 3.5 mSv per year and exceeds the 1 mSv per year limit for ordinary citizens set by Japanese law. These freelance monitoring efforts are starting to have an impact and the Tokyo Metropolitan Government added an additional 100 monitoring stations (Normile 2011).

At Fukushima, many public sector workers have experienced occupational exposure to radiation (Yokogawa et al., 2011). The area around the stricken power plant is considered an emergency evacuation preparation zone with a radius of 30 km, an inner no-entry zone of 20 km and a planned evacuation zone. The evacuation zone comes under the jurisdiction of local governments of one prefecture, four cities, six towns and three villages, with a combined work force of about 34,000 employees, These workers oversee evacuation and temporary return of residents, searches for bodies and surveying debris. They are joined in the designated zone by employees of various ministries and emergency agencies, including defense, fire and police. Administrative measures to protect public sector employees are under active consideration (Yokogawa et al., 2011).

Recovery workers at Fukushima remain a major concern because they are exposed to gamma radiation during their work shifts, with potential life-threatening damage to radiosensitive bone marrow (hematopoietic system damage) over time (Scott 2011). A proposed model concludes that life-threatening damage could be avoided if the whole body effective dose is less than the annual effective dose limit of 250 mSv (Scott 2011). Plant workers will now be monitored for long-term health effects (Butler 2011). The Tokyo-based Radiation Effects Association already has an ongoing study of the health of Japanese nuclear-power workers and new dosimetric data for Fukushima workers will be merged into that study. However, the Radiation Effects Research Foundation (RERF), based in Hiroshima and Nagasaki—which is responsible for radiation epidemiology studies on survivors of atomic-bomb explosions—is already initiating studies on broader Fukushima issues (Butler 2011).

The most overriding question concerns what is a safe or acceptable exposure to cesium-137, given the new environmental data on this isotope (Calabrese 2011). Despite expert governmental and non-governmental advice, the public can become confused when informed that acceptable levels of cesium-137 in Japan are more than threefold that permitted in the Ukraine. The confusion was compounded when the European Union raised acceptable levels of cesium-137 in food by 20-fold in response to Fukushima. At present, two major research gaps exist in formulating cesium-137 radioprotective criteria. The first is the absence of animal model chronic bioassays with cesium-137 wherein no reports were available regarding health effects in humans or animals that could be exclusively associated with oral exposure to cesium radioisotopes. Moreover, no reports were located in which cancer in human or animals could be associated with acute-, intermediate-, or chronic-duration oral exposure to radioactive cesium. The second research gap is the limitations of the linear low-dose-no-threshold (LNT) model that depends on estimating responses at low doses and dose rates based on evidence at high doses and dose rates; this limitation is at the core of disagreements with low dose/rate cancer risk assessments of ionizing radiations (Calabrese 2011).

People are reportedly suspicious of official assurances that the current situation will get no worse and don't trust the authorities to tell them promptly if it does (Sandman and Lanard 2011). For example, of the 80,000 residents forced to evacuate, the System for Prediction for Environmental Dose information—the process that predicts the spread of radioactive substances—was not used—leaving another 70,000 people living beyond the evacuation zone who were exposed to an estimated 10 mSv per year or ten times higher than the previous Japanese legal dose limit (Sugihara and Suda 2011). About 100,000 evacuees are still living in shelters, temporary housing units, or houses of relatives or friends, with slim prospects for the future. People in Fukushima face long-term exposure to low-dose radiation, with little evidence on the effects of such exposure to children and pregnant women. The public seemed especially hostile on learning that the government raised the radiation safety standard for children at schools from 1 mSv per year to 20 mSv per year (Sugihara and Suda 2011). The long-term effects of low-level radiation as a cause of developing cancer or other health problems, is as yet, unknown and remains a major public health concern (Bird 2011).

On July 25th 2011, perhaps to assure the population, Japan's legislature approved a supplementary budget of $1.2 billion for health care and long-term studies of the more than two million residents of Fukushima Prefecture affected by radiation from Fukushima (Normile 2011a). Authorities began distributing a 12-page questionnaire that seeks to establish precisely where every resident of the prefecture was for every hour of the 15 days following the earthquake and for select periods since then. It also asks what residents ate and drank and other questions that bear on radiation exposure. In addition, each of the prefecture's estimated 360,000 youngsters age 18 and older will get a thyroid examination. All 20,000 pregnant women in the prefecture will be examined and the health of their babies tracked. The plan also calls for medical checkups for an estimated 200,000 people evacuated from the vicinity of the reactors. And there will be mental health support for those in need. This is the planned first phase of an effort expected to last 30 years (Normile 2011a).

### 6.3.2 New Low-Dose-Radiation Risk Assessment

The current system of radiation protection for humans is based on the linear-no-threshold (LNT) risk-assessment paradigm (Scott 2008). Perceived harm to irradiated nuclear workers and the public is reflected through hypothetical increased cancers. The LNT-based system of protection uses easy measures of radiation exposure. These measures include the equivalent dose (a biological damage-potential-weighted measure) and the effective dose (equivalent dose multiplied by a tissue-specific relative sensitivity factor), as measured in sieverts (Sv) and millisieverts (mSv). Such a linear relationship, if correct, means that doubling the radiation dose doubles the risk of harm and reducing the dose reduces the risk by the same factor. Two types of radiation are usually distinguished characterizing radiation risks to humans. High linear energy transfer (LET) forms such as alpha particles, neutrons and heavy ions that produce ionization patterns when interacting with biological tissue and low LET forms, including X rays, gamma rays and beta particles that deposit far less energy when traversing a narrow thickness of tissue (Scott 2008). However, key studies on which the LNT is predicated are allegedly flawed (Calabrese et al., 2011a).

Human radiation exposures are limited for nuclear workers, the public and other groups based on limiting the effective dose. For example, the effective dose limit for nuclear workers is 50 mSv/year and for the public it is 1 mSv/year based on the U.S. Department of Energy and the U.S. Nuclear Regulatory Commission regulatory policies (Metting 2005). However, low doses are often delivered at low rates and a correction is made for a reduction in harm after low-rate exposure as compared to high-rate exposure (Mitchel 2006, 2007). For low doses and dose rates, a low dose and dose-rate effectiveness factor is used to reduce the slope of the cancer risk curve by a fixed amount, usually by a factor of 2. With the LNT framework, reducing the effective dose by a factor of 2 has the same effect. By using the LNT-based approach for low-dose, low-dose rate risk assessment, one essentially dismisses the possibility of radiation-induced protective effects (hormesis) as the dose-response curve slope is constrained to be positive (Mitchel 2006, 2007).

With hormesis, low doses of radiation protect against cancer, leading to a negative slope in the low-dose region for the dose-response curve. High doses, however, inhibit protection causing risk to increase as dose increases (Calabrese and Baldwin 2001; Calabrese 2004, 2005; Calabrese et al., 2006; Calabrese et al., 2011a; Tucker 2012).

Three classes of radiation hormesis are distinguished (Calabrese et al., 2007): (1) *Radiation conditioning hormesis*: this form of hormesis occurs when small or moderate radiation doses administered at a low rate activates protective processes that suppress harm from subsequent damaging large radiation doses. (2) *Radiation hormesis:* a small or moderate radiation dose given at a low rate activates protective processes and reduces the level of biological harm to below the spontaneous level. (3) *Radiation post-exposure conditioning hormesis:* damage normally caused by a large radiation dose or large dose of some other agent is reduced as a result of a subsequent exposure to a small or moderate radiation dose delivered at a low rate.

The indicated radiation-associated hormesis publications and others collectively demonstrate that low doses and dose rates of low linear energy transfer: (1) activate protective programmed cell death (apoptosis) signaling pathways and stimulate immunity; (2) protect against spontaneous chromosomal damage, mutations, neoplastic transformation and cancer; and (3) protect against high dose chemical- and radiation-induced cancer.

## 6.4 Soil and Food

### *6.4.1 Japan*

Soil contamination by atmospheric fallout is due to the dry deposition and wet deposition (rain out and washout) of dust and aerosols onto the underlying surface (Grubich 2012). Regardless of surface area, the spatial distribution of cesium-137 and other soil radio contaminants was described by a log normal distribution with similar patterns at Chernobyl and Fukushima (Grubich 2012).

Results of daily and monthly deposition samples taken since March 11th, 2011 show that the Fukushima-derived radioactive cloud affected all of the Japanese land area, especially the central and eastern part of Honshu Island and also the western North Pacific Ocean (Hirose 2011). Geographical distribution by prevailing winds of radioactive iodine, cesium and tellurium in surface soils of central-east Japan was determined by gamma-ray spectrometry (Kinoshita et al., 2011). Radioactive material transported on March 15th was the major contributor to contamination in Fukushima prefecture whereas the radioactive material transported on March 21st was the major source in Ibaraki, Tochigi, Saitama and Chiba prefectures and in Tokyo (Kinoshita et al., 2011). Cesium-137 was detected in soil from four Japanese prefectures on April 26-27th between 4 and 29 $Bq/m^2$; for iodine-131, at one prefecture, the maximum value recorded was 3.3 $Bq/m^2$ (IAEA 2011). Cesium-137 strongly contaminated the soils of large areas of eastern and northeastern Japan, compared to western

Japan which was sheltered by mountains (Yasunari et al., 2011). Soils around Fukushima and neighboring prefectures were contaminated with depositions of 100,000 and 10,000 MBq/km$^2$ (1 MBq = 1 million Bq), respectively. Total cesium-137 depositions of the Japanese Islands was estimated to be more than 5.6 PBq (1 PBq = $10^{15}$ Bq; and for the surrounding ocean it was 1.0 PBq (Yasunari et al., 2011).

The deposition velocities (rate of deposition rate to concentration) of cesium radionuclides and iodine-131 were detectably different (Amano et al., 2011) Attempts to extract cesium-134, cesium-137 and iodine-131 with water from radioactive Fukushima surface soils where rapeseed and wheat were grown resulted in 20% removal but the rest of the nuclides were not amenable to further water extraction (Nogawa et al., 2011). In the case of paddy soil, another 20% was removed with treatment by a solution of water containing potassium iodide, cesium iodide, fertilizer, calcium hydroxide and cement; the extraction rate was the same as that of water alone (Nogawa et al., 2011). In an undisturbed paddy field before plowing in Fukushima prefecture on May 24th, 2011, 88% of cesium 134 + 137 were detected in the upper 3 cm, 8% between 3 and 5 cm and the remainder between 10 and 15 cm (Shiozawa et al., 2011). Collective velocity of cesium was about 1/10 of water due to sorption of cesium on soil, indicating that sorption equilibrium is not the dominant process for predicting cesium movement is soils (Shiozawa et al., 2011). Soil samples were collected 20 km south of Fukushima and analyzed for radioactivity (Tagami et al., 2011). Concentrations and activity ratios of iodine-131, cesium-134, cesium-136, cesium-137 and tellurium-129m were measurable, but only trace amounts of niobium-95, silver-110m and lanthanum-140 were detected and these were too low to provide accurate concentrations. Not detected were zirconium-95, ruthenium-103, ruthenium-106 and barium-140: these were present in Chernobyl fallout. Authors conclude that noble gasses and volatile radionuclides predominated in the releases from from the Fukushima nuclear reactor to the terrestrial environment (Tagami et al., 2011a).

Radioactive contaminants in soil samples were measured at four locations northwest of the Fukushima reactors, at four schools and in four cities including Fukushima City (Endo et al., 2011). Isotopes measured at all locations included tellurium-129, tellurium-129m, tellurium-132, iodine-131, cesium-134, cesium-136, cesium-137, barium-140 and lanthanum-140. The highest soil depositions were at locations northwest of the reactors. From these soil deposition data, variations in dose rates over time and the cumulative external doses at 3 months and one year after deposition were estimated. At locations northwest of the Fukushima reactors, the external dose rate at three months was 4.8-98 *u*Sv/hour and the cumulative dose for one year was 51–1,000 mSv; the highest values were at Futaba and Yamada. At the schools and the four urban cities, the external dose rate at three months after deposition ranged from 0.03 to 3.8 *u*Sv/hour and for cumulative doses for one year it ranged from 3 to 40 mSv. Authors conclude that radionuclide deposition in soils are important in calculating external dose rates and cumulative doses and these data should be considered when formulating countermeasures and remediation actions (Endo et al., 2011). Radioactive soil has been removed from around schools and other institutions near Fukushima, but it remains in large mounds or has been buried at shallow depths; a remaining hundreds of thousands of tons of radioactive soil will have to be dealt with (Brumfiel and Cyranoski 2011).

Leaves from cabbages and potatoes grown near Tokyo about 230 km from the nuclear plant 40-47 days after the March 11th accident were analyzed for cesium-134 and cesium-137 (Oshita et al., 2011). Total radioactivity in both plants was less than 9 Bq/kg, well below the regulated value of 500 Bq/kg for these isotopes. Soil radioactivity was about 130 Bq/kg, less than that of the naturally-occurring potassium-40 of about 290 Bq/kg (Oshita et al 2011). Maximum radioactivity levels for iodine and cesium in leafy vegetables reported in April 2011 were 54,100 Bq/kg for iodine-131 and 82,000 Bq/kg of cesium-134 plus cesium-137 (Hamada and Ogino 2011). Some animal food products were also contaminated with iodine-131 (mostly milk) and to a lesser extent with cesium-134 and cesium-137 (Beresford and Howard 2011). Elevated levels of radioactivity were measured in local milk and agricultural products as early as March 21st (Table 6.1). The Japanese government promptly removed contaminated food—including beef and spinach—from the shelves of grocers, banned veal and milk from Fukushima and environs and imposed tapwater constraints when necessary (Osnos 2011). In early April, near the Fukushima reactors, fish (sand lance) contained up to 100,000 Bq/kg and some species of algae up to 100 million Bq/kg (Schiermeier 2011). The current Japanese legal limit for radioactivity in fish for human consumption is 500 Bq/kg for cesium-137 and 2,000 Bq/kg for iodine-131. On May 9th, tea picked in Tokyo contained 2,700 Bq of cesium-134 (Table 5.6) and on October 13th, strontium-90 was found in a soil sample from Yokohama, the first time that strontium-90 was detected more than 90 km from the plant (Harlan 2011c).

**Table 6.1. Radiation levels measured in soil and food from Fukushima environs (Wikipedia 2011).**

| *Date* | *Measured radiation* |
|---|---|
| March 21-22, 2011 | Elevated levels found in milk and agricultural products. Some products prohibited for human consumption. Elevated levels from solid waste areas, possibly plutonium-238, plutonium-239 and plutonium-240 |
| March 30–May 5 | Elevated levels of strontium isotopes 22 to 62 km distant. Maximum levels of 250 Bq/kg of strontium-90 and 1,250 Bq/kg of strontium-89. |
| May 9 | Tea picked in Tokyo contained up to 2,700 Bq/kg of cesium-134. |
| May 18 | No elevated levels of selected elements including uranium-234, uranium-235 and uranium-238 |
| May 29 | Seabed radiation levels elevated hundreds of times above background up to 36 km distant; specifically, a 300 km stretch from Miyagi Prefecture to Chiba Prefecture |
| April 4–July 4, 2011 | Elevated levels of radiocesium-137 found in small fish caught off coast. Radiocesium is expected to enter the Japanese seafood supply and is predicted to reach the U.S. west coast in 5 years. |

In June 2011, radioactivity of iodine-131, cesium-134 and cesium-37 in milk and soil, pasture (Italian ryegrass) and well water supplied to cows was measured in Ibaraki prefecture, about 130 km south of Fukushima nuclear power plant 3 (Hashimoto et al., 2011). Radionuclides in meadow grass were promptly transferred to milk; however levels were lower than the provisional government regulated value of 300 Bq/kg for radioactive iodine and 200 Bq/kg radiocesium (Hashimoto et al., 2011). Food monitoring data were reported by the Japanese Ministry of Health, Labour and Welfare on April 27th for a total of 129 samples collected from ten prefectures between April 21st and 27th (IAEA 2011). Analytical results for 125 of the 129 samples of vegetables, mushrooms, strawberries, pork, seafood and raw or fresh milk had iodine-131 and cesium 134 and cesium-137 concentrations that were either not detected or were below regulation values set by the Japanese authorities. In Fukushima prefecture, two samples of spinach collected April 24th and 25th and two samples of seafood (sand lance) collected on April 26th were above the regulation values of cesium-134 and cesium-137. On April 27th, restrictions were lifted on the distribution of spinach in Tochigi prefecture. In Fukushima prefecture, restrictions were lifted on the distribution and consumption of head-type leafy vegetables from 17 locations and flower head brassicas from nine locations. (IAEA 2011). Radioactivity imaging and measurement of radiocesium isotopes in a wheat plant grown in Fukushima in May 2011 showed that activity was about 100 times higher in withered leaves (300 Bq/kg) than new leaves (3 Bq/kg), suggesting that radiocesium strongly adheres to leaves (Tanoi et al., 2011).

There are several cases of inadequate monitoring of rice, one of the more egregious being rice grown in Onami, about 50 km northwest of Fukushima (Fackler 2012). Government inspectors declared Onami's rice safe for consumption after testing just two of its 154 rice farms. A skeptical farmer in Onami who wanted to be sure that his rice was safe for a visiting grandson had his crop tested, only to find that it contained levels of cesium that exceeded the government's safety limit. In the weeks that followed, at least a dozen farmers found unsafe levels of cesium in their crops. The ensuing panic forced an intervention by the Japanese government with promises to test more than 25,000 rice farms in eastern Fukushima where the plant is located. Many citizens, including many experts, came to believe that officials have understated or even covered up the true extent of the public health risk in order to limit both the economic damage and the size of potential compensation payments One consequence of perceived government inefficiency was the formation of a testing center by farmers (with a $40,000 testing device contributed by a non-governmental group) that listed radiation levels for all of their crops with the results posted online. One farmer destroyed his entire crop of 110,000 mushrooms after tests revealed unacceptably high radiation. So far, TEPCO has offered full compensation only to farmers in the zones that were evacuated and a larger area to the northwest where winds carried much of the fallout. Farming officials say they have too few radiation-detecting machines to test every product from every farm; there are only a few dozen machines in all of Fukushima Prefecture, a region about the size of Connecticut, with 110,000 farms (Fackler 2012).

Radioactivity in Chiba tap water was detected several days after radioactive fallout in the area (Amano et al., 2011). Radiation doses were estimated from external radiation and internal radiation by inhalation and ingestion of tap water for people living outdoors in the Chiba metropolitan area following the Fukushima accident (Amano et al., 2011). Since April 1st only one restriction remained: the consumption of drinking water relating to iodine-131 (with a limit of

100 Bq/L) in one village in Fukushima prefecture and only for infants (IAEA 2011). Attempts to remove iodine-131 from tapwater by boiling for as much as 30 minutes were unsuccessful; indeed, even higher iodine-31 concentrations may result from the liquid-volume reduction accompanying this process (Tagami and Uchida 2011).

When radiation readings from water monitors around the leaking Fukushima Daiichi reactors spiked at 7.5 million times Japan's legal limit for radioisotopes in public water in April 2011, the Japanese government—for the first time in history—set a limit on radiation in seafood and began screening fish (Reardon 2011). Screening was conducted by the National Research Institute of Fisheries Science in Japan and they found elevated radiocesium levels in a few fish. Results were largely discounted because research from weapons testing and from the Chernobyl accident ascertained that large fish that accumulate cesium-137 excrete it over time (Reardon 2011).

### *6.4.2 Europe and North America*

Cow's milk in northwest Germany between March and May 2011, contained traces of radioiodine and radiocesium, as did the grass that they were fed (Pittauerova et al., 2011). In France, samples of grass, milk and vegetables contained only traces of iodine-131 and radiocesium isotopes from Fukushima. In northwest Germany, fallout was found in soil and river sediments collected between March and May 2011 (Pittauerova et al., 2011). Rainwater and milk from cows, sheep and goats collected in Romania on March 27th, 2011 contained traces of iodine-131, with maximum values measured of 0.75 Bq/L in rain and 5.2 Bq/L in milk (Margineanu et al., 2011).

From March 24th to 30th, Fukushima-derived radioiodine-131 was detected in milk from the United States (Parache et al., 2011; Wikipedia 2011). The apparent dry deposition velocity of iodine on grass ranged between 1,000 and 5,000 meters per second from site to site. In addition, the grass to milk transfer factors ranged from 2,800 for goat's milk to 36,000 for cow's milk (Parache et al., 2011).

Between April 3rd and April 26th, 2011, elevated levels of iodine-131 and cesium-137 were detected in rainwater from the Greater Sudbury area of eastern Canada, when compared to a reference sample of ice-water (Cleveland et al., 2012). These elevated activities are ascribed to the accident at the Fukushima Daiichi nuclear reactor that occurred on March 11th, 2011. The activity levels observed at no time presented health concerns (Cleveland et al., 2012).

### 6.4.3 Others

India and China have banned imports of food products from certain areas of Japan (Reardon 2011).

## 6.5 Aquatic Monitoring Program

Marine monitoring was conducted near the discharge areas of the Fukushima nuclear plant by TEPCO and offshore by MEXT (IAEA 2011). Marine contamination occurred from aerial deposition and by discharges and outflow of contaminated water. Samples from offshore stations were measured for ambient dose rate in air above the sea and in ambient dust above the sea. In addition, samples of surface seawater and seawater collected 10 m above the sea bottom were analyzed. As of April 26th and 27th, analysis of almost all sampling positions has shown the following: a general decreasing trend in radionuclide concentrations over time; samples from coastal locations show higher concentrations than offshore samples; and radionuclides of iodine-131, cesium-137 and cesium-134 were still detected in most seawater samples, but no longer for some of the off-shore positions (IAEA 2011). On April 6th, the measured cesium-137 concentration in a seawater sample near the Fukushima reactor was 6.8 kBq/L = $6.8 \times 10^4$ Bq/L (Tsumune et al., 2011). Two major pathways from the accident site to the ocean were direct release of high radioactive liquid wastes to the ocean and the deposition of airborne radioactivity to the ocean surface. Authors concluded that direct release from the site was the dominant mechanism (Tsumune et al., 2011).

Similar maximum levels of fallout radionuclides (cesium-134, cesium-137, iodine-131) were detected in water samples from the United States, Greece and Krasnoyarsk (Russia, situated in central Asia), demonstrating the high-velocity movements from Fukushima and the global effects of this accident such as those caused by the Chernobyl accident (Bolsunovsky and Dementyev 2011).

Seawater from 10 sites along the coastline 250–450 km north of the Fukushima reactors were collected in May-June 2011 and analyzed for cesium-134 and cesium-137 using gamma spectrometry (Inoue et al., 2011). Cesium-134 and cesium-137 activities in May were 2-3 mBq/L and 2.5-4 mBq/L respectively. By June, these values had decreased by 25 to 45% per month for cesium-134 and by 5 to 30% per month for cesium-137. Authors aver that surface infusion of these isotopes into the sea by atmospheric transport from Fukushima and their subsequent reduction by water migration to off-shore and deeper regions could account for the decreases (Inoue et al., 2011). More research on cesium-137 and cesium-134 levels in aerosols is recommended for their fate in lake sediments of Japan (Kanai 2011)

In Korea, Fukushima-derived iodine-131 and cesium-134 and cesium-137 were detected in dry and wet deposition samples on April 7, 2011; however there was no measurable increase of these radionuclides in seawater or marine biota (Kim et al., 2011).

On June 25th, 2012, the first seafood caught off the Fukushima coastline went on sale since March 2011, but the offerings were limited to giant octopus and whelks (marine snails); these were chosen because they contained no detectable radiocesium (Associated Press 2012). But flounder, sea bass and other fish were still prohibited because of unacceptably high levels of radiation. (Associated Press 2012).

## LITERATURE CITED

Amano, H., M. Akiyama, B. Chunlei, T. Kawamura, T. Kishimoto, T. Kuroda, T. Muroi, T. Odaira, Y. Ohta, K. Takeda, Y. Watanabe and T. Morimoto. 2011. Radiation measurements in the Chiba metropolitan area and radiological aspects of fallout from the Fukushima Daiichi nuclear power plants accident, *Journal of Environmental Radioactivity*, doi:10.1016/j.jenvrad,2011.10.019.

Associated Press. 2012. Japan sells first fish caught off Fukushima coast since nuclear crisis, http://www.dailynews.com/news/ci-20932786/japan-sells-first fish-caught-off Fukushima-coast?source=rss.

Beresford, N.A. and B.J. Howard. 2011, An overview of the transfer of radionuclides to farm animals and potential countermeasures of relevance to Fukushima releases, *Integrated Environmental Assessment and Management*. **7(3)Special Issue:**382–384.

Biegalski, S.R., T.W. Bowyer, P.W. Eslinger, J.A. Friese, L.R. Greenwood, D.A. Haas, J.C. Hayes, I. Hoffman, M. Keillor, H.S. Miley and M. Moring. 2011. Analysis of data from sensitive U.S. monitoring stations for he Fukushima Daiichi nuclear reactor accident, *Journal of Environmental Radioactivity*, doi:10.1016/j.jenvrad.2011.11.007.

Bird, W.A. 2011. Fukushima health study launched, *Environmental Health Perspectives*. **119(10):**a428-a429.

Bolsunovsky, A. and D. Dementyev. 2011. Evidence of the radioactiv fallout in the center of Asia (Russia) following the Fukushima nuclear accident, *Journal of Environmental Radioactivity*. **102(11):**1062–1064.

Bowyer, T.W., S.R. Biegalski, M. Cooper, P.W. Eslinger, D. Haas, J.C. Hayes, H.S. Miley, D.J. Strom and V. Woods. 2011, Elevated radioxenon detected remotely following the Fukushima nuclear accident, *Journal of Environmental Radioactivity*. **102:**681–687.

Brumfiel, G. and D. Cyranoski. 2011. Fukushima deep in hot water. *Nature.* **474:**135–136.

Butler, D. 2011. Fukushima health risks scrutinized, *Nature.* **472:**13-14.

Calabrese, E.J. 2004. Hormesis: From marginalization to mainstream: A case for hormesis as the default dose-response model in risk assessment, *Toxicology and Applied Pharmacology.* **197:**125–136.

Calabrese. E.J. 2005. Paradigm lost, paradigm found: The re-emergence of hormesis as a fundamental dose response model in toxicological sciences, *Environmental Pollution.* **138:**379–412.

Calabrese, E.J. 2011. Improving the scientific foundations for estimating health risks from Fukushima, *Proceedings of the National Academy of Sciences of the United States.* **108(49):**19447–19448.

Calabrese, E.J. 2011a. Key studies used to support cancer risk assessment questioned, *Environmental and Molecular Mutagenesis.* **52(8):**595–606.

Calabrese, E.J., K.A. Bachman, A.J. Bailer, P.M. Bolger, J. Borak, L. Cai, N, Cedergreen, M.G. Cherian, C.C. Chiueh, T.W. Clarkson, R.R. Cook, D.M Diamond, D.J. Doolittle, M.A. Dorato, S.O. Duke, L. Feinendegen, D.E. Gardner, R.W. Hart, K.L. Hastings, A.W. Hayes, G.R. Hofmann, J.A. Ives, Z. Jaworski, T.E. Johnson, W.B. Jones, N.E. Kaminski, J.G. Keller, J.E. Klaunig, T.B. Knudsen, W.J. Kozumbo, T. Lettieri, S.Z. Liu, A. Maisseu, K.I. Maynard, E.J. Masoro, R.O. McClellan, H.M. Mehendale, C. Mothersill, D.B. Newlin, H.N. Nigg, F.W. Oehme, R.E. Phalen, M.A. Philbert, S.I. Rattan, J.E. Riviere, J. Rodricks, R.M. Sapolsky, B.R. Scott, C. Seymour, D.A. Sinclair, J. Smith-Sonneborn, E.T. Snow, L. Spear, D.E. Stevenson, Y. Thomas, M. Tubiana, G.M. Williams and M.P. Mattson. 2007. Biological stress response terminology: integrating the concepts of adaptive response and preconditioning stress within a hormetic dose-response framework, *Toxicology and Applied Pharmacology.* **222:**122–128.

Calabrese E.J. and L.A. Baldwin. 2001. Hormesis: U-shaped dose-responses and their centrality in toxicology, *Trends in Pharmacological Science.* **22:**285–291.

Calabrese, E.J., J.W. Staudenmayer, E.J. Stanek III and G.R. Hoffmann. 2006. Hormesis outperforms threshold model in National Cancer Institute antitumor drug screening database, *Toxicological Science.* **94(2):**368–378.

Cleveland. B.T., F.A. Duncan, I.T. Lawson, N.J.T. Smith and E. Vazquez-Jauregui. 2012. Activities of *γ-ray* emitting isotopes in rainwater from Greater Sudbury, Canada following the Fukushima incident, *Preprint submitted to Journal of Environmental Radioactivity,* January 20, 2012, 4 p.

Endo, S., S. Kimura, T. Takatsuji, K. Nanasawa, T. Imanaka and K. Shizuma. 2011. Measurement of soil contamination by radionuclides due to the

Fukushima Daiichi nuclear power plant accident and associate estimated cumulative external dose estimation, , doi:10.1016/j.jenvrad..2011.11.006

Fackler, M. 2012. Japanese struggle to protect their food supply, *New York Times (newspaper)*, January 22nd, A6.

Grubich, A.O. 2012. Multifractal structure of the $^{137}Cs$ fallout at small spatial scales, *Journal of Environmental Radioactivity*. **107:**51–55.

Gudelis, A., R. Druteikiene, G. Lujaniene, E. Macieka, A. Plukis and V. Remeikis. 2012. Radionuclides in the ground-level atmosphere in Vilnius, Lithuania, in March 2011, detected by gamma-ray spectrometry, *Journal of Environmental Radioactivity*. **109:**13–18.

Hamada, N. and H. Ogino. 2011. Food safety regulations: What we learned from the Fukushima nuclear accident, *Journal of Environmental Radioactivity*, doi:1016/j.envrad.2011.08.008.

Harlan, C. 2011c. Small hot spots in Tokyo heighten worry about radiation spread, *Washington Post (newspaper)*, October 14th, A 7.

Hashimoto, K.E.N., K. Tanoi, K. Sakurai, T. Iimoto, N. Nogawa, S. Higaki, N. Kosaka, T. Takahashi, Y. Enomoto, I. Onoyama, J.U.N. You Li, N. Manabe and T.M. Nakanishi. 2011. The radioactivity measurement of milk from the the cow supplied with the meadow grass grown in Ibaraki-prefecture, after the nuclear plant accident, *Radioisotopes*. **60(8):**335–338.

Hirose, K. 2011. Fukushima Daiichi nuclear power plamt accident: Summary of regional radioactive deposition monitoring results, *Journal of Environmental Radioactivity*, doi:10.1016/j.jenvrad.2011.09.003.

IAEA (International Atomic Energy Agency). 2011. Fukushima nuclear accident update log, iaea.org/...tsunamiupdate01.html

Inoue, M., H, Kofuji, Y. Hamajima, S. Nagao, K. Yoshida and M. Yamamoto. 2011. $^{134}Cs$ and $^{137}Cs$ activities in coastal seawater along northern Sanriku and Tsugaru Strait, northeastern Japan, after Fukushima Dai-ichi nuclear power plant accident, doi:10.1016/j.jenvrad.2011.09.012.

Kanai, Y. 2011. Monitoring of aerosols in Tsukuba after Fukushima nuclear power plant incident in 2011, *Journal of Environmental Radioactivity*, doi.org/10.1016/j.jenvrad.2011.10.011.

Kim, C.K., J.I. Byun, J.S. Chae, H.Y. Choi, S.W. Choi, D.J. Kim, Y.J. Kim, D.M. Lee, W.J. Park, S.A. Yim and J.Y. Yun. 2011. Radiological impact in Korea following the Fukushima nuclear accident, doi:10.1016/j.jenvrad.2011.10.018

Kinoshita, N., K. Seki, K. Sasa, J.I. Kitagawa, S. Karashi, T. Nishimura, Y.S. Wong. Y. Satou, K. Handa, T. Takahashi, M. Sato and T. Yamagata. 2011. Assessment of individual radionuclide distributions from the Fukushima nuclear accident covering central-east Japan, *Proceedings of the National Academy of Sciences of the United States*. **108(49):**19526–19529.

Leelossy, A., R. Meszaros and I. Lagzi. 2011. Short and long term dispersion patterns of radionuclide in the atmosphere around the Fukushima nuclear power plant, *Journal of Environmental Radioactivity*. **101(12):**1117–1121.

Leon, J.D., D.A. Jaffe, J. Kaspar, A. Knecht, M.L. Miller, R.G.H. Robertson and A.G. Schubert. 2011. Arrival time and magnitude of airborne fission products from Fukushima, Japan, reactor incident as measured in Seattle, WA, USA, *Journal of Environmental Radioactivity.* **102(11):**1032–1038.

Margineanu, R., B. Mitrica, A.N.A. Apostu and C. Gomoiu. 2011. Traces of radioactive 131I in rainwater and milk samples in Romania, *Environmental Research Letters,* 6. 034011, doi:10.1088/1748-9326/6/3/034011

O. Masson, A. Baeza, J. Bieringer, K. Brudecki, S. Bucci, M. Cappai, F.P. Carvalho, O. Connan, C. Cosma, A. Dalheimer, D. Didier, G. Depuydt, L.E. De Geer, A. De Vismes, L. Gini, F. Groppi, K. Gudnason, R. Gurriaran, D. Hainz, Ó. Halldórsson, D. Hammond, O. Hanley, K. Holeý, Zs. Homoki, A. Ioannidou, K. Isajenko, M. Jankovic, C. Katzlberger, M. Kettunen, R. Kierepko, R. Kontro, P.J.M. Kwakman, M. Lecomte, L. Leon Vintro, A.-P. Leppánen, B. Lind, G. Lujaniene, P. Mc Ginnity, C. Mc Mahon, H. Malá, S. Manenti, M. Manolopoulou, A. Mattila, A. Mauring, J.W. Mietelski, B. Møller, S.P. Nielsen, J. Nikolic, R.M.W. Overwater, S.E. Pálsson, C. Papastefanou, I. Penev, M.K. Pham, P.P. Povinec, H. Ramebäck, M.C. Reis, W. Ringer, A. Rodriguez, P. Rulík, P.R.J. Saey, V. Samsonov, C. Schlosser, G. Sgorbati, B. V. Silobritiene, C. Söderström, R. Sogni, L. Solier, M. Sonck, G. Steinhauser, T. Steinkopff, P. Steinmann, S. Stoulos, I. Sýkora, D. Todorovic, N. Tooloutalaie, L. Tositti, J. Tschiersch, A. Ugron, E. Vagena, A. Vargas, H. Wershofen and O. Zhukova. 2011. Tracking of airborne radionuclides from the damaged Fukushima Dai-ichii nuclear reactors by European networks, *Environmental Science & Technology.* **45(18):**7670–7677.

Matanle, P. 2011. The Great East Japan earthquake, tsunami and nuclear meltdown: Towards the (re)construction of a safe, sustainable and compassionate society in Japan's shrinking regions, *Local Environment.* **16(9):**823–847.

Metting, M. 2005. *Ionizing radiation dose ranges chart,* Available from U.S. Department of Energy, Office of Science, Washington, D.C., 20005.

Mitchel, R.E.J. 2006. Cancer and low dose responses *in vivo*: Implications for radiation protection, *Proceedings of the 15th Pacific Basin Nuclear Conference,* October 15–20, Sydney, Australia.

Mitchel, R.E.J. 2007. Low doses of radiation reduce risk *in vivo, Dose Response.* **5(1):**1–10.

Monzen, S., M. Hosoda, S. Tokonami, M. Osanai, H. Hoshino, Y. Hosokawa, M.A. Yoshida, M. Yamada Y. Asari, K. Satoh and I. Kashiwakura. 2011. Individual radiation exposure dose due to support activities at safe shelters in Fukushima prefecture, *PLoS ONE.* **6(11):**e27761, doi:19.1371/journal.pone.0027761.

Nogawa, N., K,E.N. Hashimoto, K. Tanoi, T.M. Nakanishi, N. Nihei and Y. Ono. 2011. Extraction of Cs-137, Cs-134 and I-131 from radioactive soils in Fukushima, *Radioisotopes.* **60(8):**311–315.

Norman. E.B., C.T. Angell and P.A. Chodash. 2011. Observations of fallout from the Fukushima reactor accident in San Francisco Bay area rainwater, *Plos One*. **6(9)**:article e24330..

Normile, D. 2011. Citizens find radiation far from Fukushima, *Science*. **332(6036)**:1368.

Normile, D. 2011a, Fukushima begins 30-year odyssey in radiation health, *Science*. **333(6043)**:684–685,

Oshita, S., Y. Kawagoe, E. Yasunaga, D. Takata, T.M. Nakanishi, K. Tanoi, Y. Makino and H. Sasaki. 2011. Radioactivity measurement of soil and vegetables contaminated from low level radioactive fall out arised (sic) from Fukushima Daiichi nuclear accident—a study on Institute for Sustainable Agro-ecosystem Services Graduate School of Agricultural and Life Sciences, the University of Tokyo-, *Radioisotopes*. **60(8)**:329–333.

Osnos, E. 2011. Letter from Fukushima. The fallout. *The New Yorker*, October 17, pp. 46–61.

Parache, V., L. Pourcelot, S. Roussel-Debet, D. Orjollet, F. Leblanc, C. Soria, R. Gurriaran, P.H. Renaud and O. Masson. 2011. Transfer of $^{131}$I from Fukushima to the vegetation and milk in France, *Environmental Science & Technology*. **45(23)**:9998–10003.

Pittauerova, D., B. Hettwig and H.W. Fischer. 2011. Fukushima fallout in Northwest German environmental media, *Journal of Environmental Radioactivity*, **102(9)**:877–880.

Potiriadis, C., M. Kolovou, A. Clouvas and S. Xanthos. 2011. Environmental radioactivity measurements in Greece following the Fukushima Daiichi nuclear accident, *Radiation Protection Dosimetry*, doi:10.1093/rpd/ner423.

Reardon, S. 2011. Fukushima radiation creates unique test of marine life's hardiness, *Science*. **332(6027)**:292.

Ringer, W., J. Klimstein and M. Bernreiter. 2011. Long-term series and activity size distribution of beryllium-7 in the atmospheric environment time series and activity size distribution of iodine-131 in Austria after the Fukushima NPP accident, *Radioprotection*. **46(6)**:S7–S10.

Sandman, P.M. and J. Lanard. 2011. It is rational to doubt Fukushima reports, *Nature*. **473**:31.

Schiermeier, Q. 2011. Radiation release will hit marine life, *Nature*. **472**:145–146

Scott, B.R. 2008. It's time for a new low-dose-radiation risk assessment paradigm—one that acknowledges hormesis, *Dose Response*. **6(4)**:433–451.

Scott, B.R. 2011. Assessing potential radiological harm to Fukushima recovery workers, *Dose-response*. **9(3)**:301–312,

Shiozawa, S. H.O., K. Tanoi, K. Nemoto, S, Yoshida, K, Nishida, K.E.N. Hashimoto, K. Sakurai, T.N. Nakanishi, N. Nihei and Y. Ono. 2011. Vertical concentration profiles of radioactive caesium and convective velocity in soil in a paddy field in Fukushima, *Radioisotopes*. **60(8)**:323–328.

Sinclair, L.E., H.C.J. Seyward, R. Fortin, J.M. Carson, P.R.B. Saull, M.J. Coyle, R.A. Van Brabant, J.L. Buckle, S.M. Desjardins and R. M Hall. 2011. Aerial measurement of radioxenon concentration off the west coast of Vancouver Island following the Fukushima reactor accident, *Journal of Environmental Radioactivity.* **102(11):**1018–1023.

Stohl, A., P. Seibert, G. Wotawa, D. Arnold, J.F. Burkhart, S. Eckhardt, C. Tapia, A. Vargas and T.J. Yasunari. 2011. Xenon-133 and caesium-137 releases into the atmosphere from the Fukushima Dai-ichi nuclear power plant: Determination of the source term, atmospheric dispersion and deposition, *Atmospheric Chemistry and Physics Discussions.* **11:**28319-28394, doi:10.5194/acpd-11-28319-2011,2011

Sugihara, G. and S. Suda. 2011. Need for close watch on children's health after Fukushima disaster, *The Lancet.* **378:**485-486.

Tagami, K. and S. Uchida. 2011. Can we remove iodine-131 from tap water in Japan by boiling?—Experimental testing in response to the Fukushima Daiichi nuclear power plant accident, *Chemosphere.* **84(9):**1282–1284.

Tagami, K., S. Uchida, Y. Uchihori, N. Ishii, H. Kitamura and Y. Shirakawa. 2011a. Specific activity and activity ratios of radionuclides in soil collected about 20 km from the Fukushima Daiichi nuclear power plant: radionuclide release to the south and southwest, *Science of the Total Environment.* **409(22):**4885–4888.

Tanoi, K., K.E.N. Hashimoto, K. Sakurai, N. Nihei, Y. Ono and T.M. Nakanishi. 2011. An imaging of radioactivity and determination of Cs-134 and Cs-137 in wheat tissue grown in Fukushima, *Radioisotopes.* **60(8):**317–322.

Tsumune, D., T. Tsubono, M. Aoyama and K. Hirose. 2011. Distribution of oceanic $^{137}$Cs from the Fukushima Daiichi nuclear power plant simulated numerically by a regional ocean model, *Journal of Environmental Radioactivity,* doi:1016/j.jenvrad.2011.10.007.

Tucker, W. 2012. Fukushima and the future of nuclear power, *Wall Street Journal (newspaper), )* March 6th, A 19.

Wikipedia. 2011. Timeline of the Fukushima Daiichi nuclear disaster, 21 p.

Wikipedia. 2011a. Radiation effects from Fukushima I nuclear accidents. 26 p.

Yasunari, T.J., A. Stohl, R.S. Hayano, J.F. Burkhart, S. Eckhardt and T. Yasunari. 2011. Cesium-137 deposition and contamination of Japanese soils due to the Fukushima nuclear accident, *Proceedings of the National Academy of Sciences of the United States.* **108(49):**19530–19534.

Yokogawa, T., K. Takahashi, T. Nagata, K. Mori and S. Horie. 2011. Suboptimal radiation protection for municipal employees operating in the Fukushima designated zone, *Occupational and Environmental Medicine,* doi:10.1136/oemed-2011-100436.

Zhang, W., M. Bean, M. Benotto, J. Cheung, K. Ungar and B. Ahier. 2011, Development of a new aerosol monitoring system and its application in Fukushima nuclear accident related aerosol radioactivity measurement at CTBT radionuclide station in Sidney of Canada, *Journal of Environmental Radioactivity.* **102(12):**1065–1069.

Chapter 7

# Radiation Effects

## 7.1 General

Radioactive material was released to the environment on several occasions after the tsunami struck Fukushima (Wikipedia 2011a). This occurred due to deliberate pressure-reducing venting and through accidental and controlled releases. These conditions resulted in slight amounts of radioactive contamination in the atmosphere, drinking water, milk, some crops near the reactor and in fish caught 80 km from the coast. Radioactive iodine was found in breast milk of women living east of Tokyo. Drinking water in Japan was above the limit for infants about 7 to 10 days after the accident. Several workers received dosages of more than 100 milliSieverts (mSv) and two were hospitalized with high exposures around the ankles while standing in radioactive cooling water. Water levels within units 2 and 3 were reported to be 750 to 1,000 mSv on March 27th (Wikipedia 2011a).

## 7.2 Humans

A worker standing near a single dry coolant pool at any of the Fukushima reactors would receive a fatal dose in 16 seconds (Osnos 2011). The unrecovered bodies of approximately 1,000 victims of the earthquake and tsunami within the plant's evacuation zone were believed to be contaminated with dangerous levels of radiation, but this has not been established with certainty (Wikipedia 2011a). On March 12th, the zone within 20 km of the plant was evacuated. On March

25th those living between 20 and 30 km from the site were urged to evacuate (Wikipedia 2011c).

All endocrine glands are susceptible to damage by radiation exposure, especially the pituitary, thyroid and gonads (Niazi and Niazi 2011). In addition to endocrine effects, rates of birth defects and carcinomas may also be increased in a population exposed to excessive radiation (Niazi and Niazi 2011).

The safety of workers at the Fukushima reactors was and still is, a major concern and the subject of many studies. One study proposed to protect workers against strontium-90, a fission product of uranium and plutonium, that preferentially accumulates in skeletal tissues (Dalerba 2011). Exposure to strontium-90 is associated with development of bone cancer in several animal models. Research into osteoporosis over the past 30 years has led to the development of several anti-resorptive drugs capable of inhibiting adverse effects of bone resorption and bone formation. Clinical evidence suggests that preventive administration of different types of biphosphonates is able to reduce incorporation of strontium into bone. Specifically, denosumab, an anticlonal antibody, can achieve a 6-month reversible inhibition of bone remodeling after a single subcutaneous injection. All these drugs were associated with fairly safe toxicological profiles (Dalerba 2011).

In another study, peripheral-blood stem cells, may protect rapidly-dividing cells, such as intestinal-tract and hemopoietic cells, the most vulnerable to ionizing radiation, against radiation exposure of more than 5 Gy (Tanimoto et al., 2011). On March 29th, the Japan Society for Hematopoietic Cell Transplantation said that more than 100 transplant teams were standing by to collect and store hemopoietic stem cells from peripheral blood. The European Group for Blood and Marrow Transplantation concurred and said that more than 50 hospitals in Europe have agreed to help the workers if necessary (Tanimoto et al., 2011).

There are several possible medical countermeasures to radiation exposure (Anzai et al., 2012). Medical agents are classified as absorbents or modifiers. Absorbents protect against internal exposure by preventing the accumulation or accelerating the exclusion of radionuclides, i.e., stable potassium iodide tablets can be used to prevent the absorption of radioactive iodine in the thyroid gland; Prussian blue can be used to remove cesium-137 from the human body by binding with it; diethylene-triaminepentaacetic acid known as DTPA is expected to chelate plutonium). These chemicals have been stocked for possible use after nuclear accidents and potassium iodide tablets were released to Fukushima workers. Radiation modifiers may be used to prevent or reduce injuries by modifying chemical or biological responses to radiation in the living body and are discussed in more detail by Anzai et al. (2012).

On April 22nd the Japanese prime minister stated that additional towns might be asked to evacuate (Wikipedia 2011). The government plans to build 30,000 temporary homes by the end of May and an additional 70,000 homes later. On May 12th TEPCO engineers confirmed that a meltdown had occurred; however, the Japanese government "consistently appeared to underestimate the severity of the situation". Further, the Nuclear Energy Institute, a nuclear lobbying firm, stated that the situation "is in no way alarming" (Wikipedia 2011).

Many communities in northeastern Honshu that were among the hardest hit by the earthquake and tsunami believe that life 15 years later will be even worse (Harlan 2011d). Even before the March 11th disaster, these fishing and agricultural towns along this shoreline—accounting for 2.5% of the country's economy—had a disproportionately large number of elderly, a smaller number of working adults and few small businesses. On March 11th the tsunami leveled 80% of the buildings including 92% of the small businesses, with 33% stating that they would not try to rebuild. In November, eight months later, one town's train station remains rubble, the downtown a dust bowl and the town's tax revenue zero (Harlan 2011d).

On June 3rd one worker in his 30's had received a total of 678 mSv, while another in his 40's 643 mSv (Wikipedia 2011). Before the accident the limit for emergency situation was 100 mSv but this was raised by the Japanese government to 250 mSv just after the accident. The two TEPCO workers that were on duty in the central control room of reactors 3 and 4 didn't remember whether they wore protective masks at the time of the hydrogen explosion that occurred at reactor 1 on March 12th (Wikipedia 2011).

Fear of radiation from Japan prompted a global rush for iodine pills, including the United States, Canada, Russia, Korea, China, Malaysia and Finland (Wikipedia 2011a). The United States Pentagon said that troops are receiving potassium iodide before missions to areas where possible radiation exposure is likely. Overuse of iodine tablets in response to fears about harmful radiations from Fukushima has resulted in a rash of admissions to poison centers around the world (Wikipedia 2011a).

Seven months after the triple meltdown at Fukushima Daiichi, TEPCO (which operated the facility) owed 50 billion dollars (U.S.) in compensation to the tens of thousands of residents who lived close to the reactors (Harlan 2011b). The payments could send the company into bankruptcy. At the very least, the utility will cut jobs, reduce salaries, sell its assets and perhaps raise electricity rates for its 29 million customers. With 37,000 employees earning an average salary of $100,000 (U.S.) per annum, the prognosis is grim. TEPCO, in the first few months of the crisis, squandered most of its credibility and all of its good will. Its president disappeared from public view, then resigned. It disclosed the meltdowns at the plant nearly two months after the fact, it spun false stories about the timeline of events at the facility, trying to affix blame to the government. Economists say that the company will either go bankrupt if its nuclear reactors don't start, or carry for years its debt to evacuees and lenders (Harlan 2011b).

By December 2011, an estimated 88,000 people had been evacuated from the Fukushima nuclear plant area, with many still unemployed and living in temporary housing (Fackler 2011a). When the Japanese government finally released a map of the fallout, some of the heaviest concentrations were in the village of Iitate (population 6,200), 41 km northwest of the plant. All citizens were forced to evacuate by May 1st (Fackler 2011a; Figure 7.1). Residents received the equivalent of one chest X-ray every 12 hours (Osnos 2011). The 3,000 cows in Iitate village were contaminated and subsequently killed (Osnos 2011). The 7,000 residents of Futaba were evacuated to an abandoned high school more than 160 km from home, leaving the land uninhabitable—perhaps for decades—thus joining 300,000 other refugees left homeless (Harlan 2011a). Unlike other communities abandoned because of earthquake, tsunami and radiation wherein residents dispersed to widely separated communities, the residents of Futaba proposed to relocate the entire town to a new location, perhaps in a different prefecture, thus if the land was deemed safe, the property would revert to their descendants (Harlan 2011a). Under Tokyo's reconstruction guidelines, the central government will pay to acquire land on high ground if at least five households wish to move there together (Onishi 2012). But the land must meet cost requirements established by local governments. With little flat land available, most proposed locations need authorities to buy inland mountains from individual owners and flatten them for residential use (Onishi 2012).

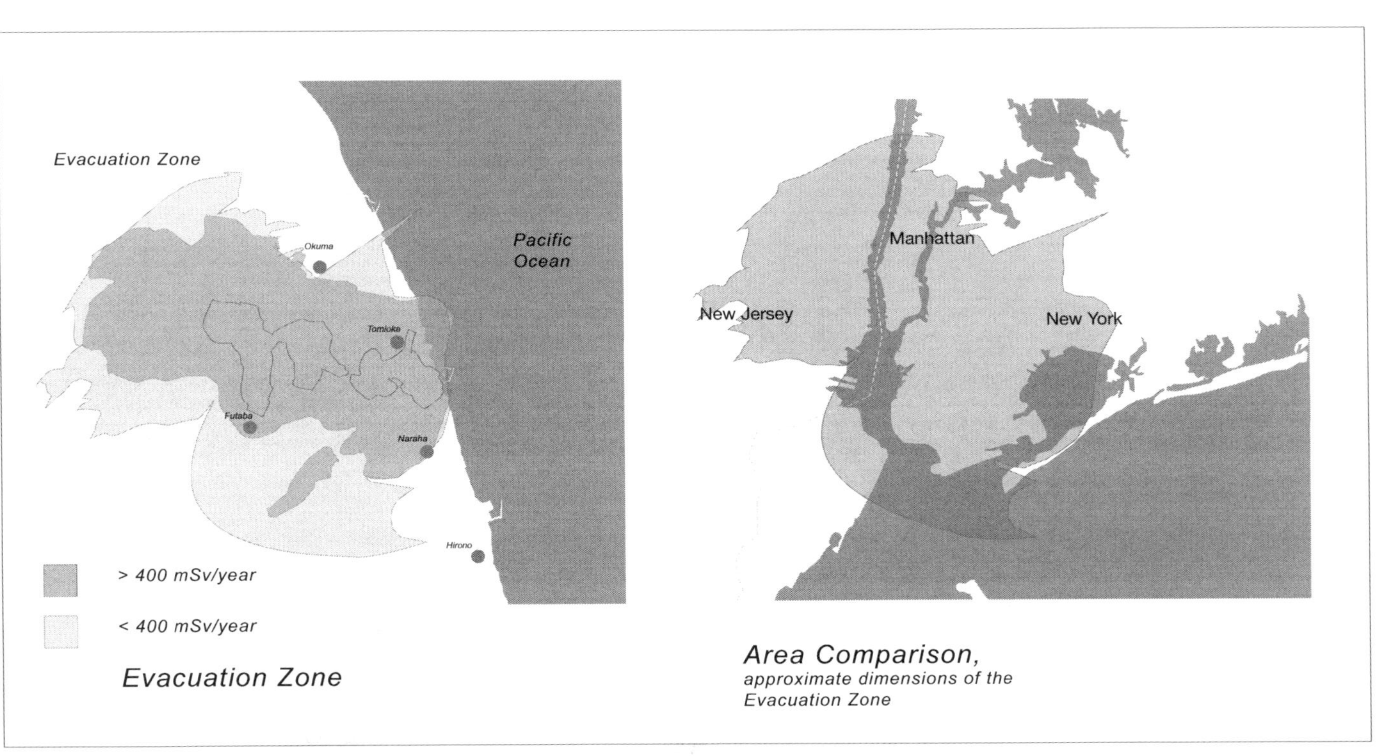


**Figure 7.1.** Fukushima Daiichi evacuation zone, with areas receiving 100 to more than 400 mSv/year highlighted and area comparison with metropolitan New York City (modified from Fackler 2011a).

*(Color image of this figure appears in the color plate section at the end of the book.)*

Until recently, Japan held that only people who experienced the atomic bomb blasts of Hiroshima and Nagasaki at close range were victims, because secondary radiation posed negligible danger (Hayashi 2011). This decision has been reversed and the number of potential victims of the World War II bombings entitled to compensation has increased by several orders of magnitude. Likewise, many potential victims of the Fukushima disaster will have received only secondary radiation from eating contaminated food or inhaling dust, for example. They too may be entitled to compensation and medical assistance for secondary radiation effects that may occur many decades after exposure to low levels of radiation. The financial implications of this decision on TEPCO and the Japanese government is currently unknown, but is expected to be significant (Hayashi 2011).

The DNA repair protein XRCC4 is an important enhancer in promoting radiation repair pathways triggered by DNA double strand breaks (Sun et al., 2011). A screen of traditional Chinese medicine databases indicates that three potent inhibitor agents exist against XRCC4. Protein-ligand interactions were focused at Lys 188 on chain A and Lys 187 on chain B and o-hydroxyphenyl and carboxyl functional groups play critical roles in stabilizing binding conformations. Based on their findings, authors concluded that potential radiotherapy enhancers are available from traditional Chinese medicine and that the key binding elements for inhibiting XRCC4 are characterized (Sun et al., 2011).

Antibiotic resistance to clinical pathogens in humans may be traced back to resistance mechanisms in environmental bacteria; factors—such as radiation—which are likely to alter bacterial resistance is of potential and eventual consequence to human pathogens (Nakanishi et al., 2012). Authors exposed four species of bacteria: the Gram-positive *Listeria innocua, Bacillus subtilis* and two species of Gram-negative bacteria *Escherichia coli, Pseudomonas aeruginosa* to levels of gamma radiation in and around Fukushima equating to 1, 10 and 100 years equivalence exposure with alteration to susceptibility to 14 antibiotics measured as the primary endpoint. The Gram-positive species were unaffected whereas both Gram-negative species became slightly more susceptible to the antibiotics tested. These findings suggest that radiation exposure will not increase the antibiotic resistance profile of these organisms and therefor not add to the global health burden of increased antibiotic resistance to human bacterial pathogens (Nakanishi et al., 2012).

As of March 4th, 2012 public opinion was running against the use of nuclear power (Iwata 2012a). A survey conducted by national broadcaster NHK found that 51% of respondents said that they wanted less reliance on nuclear-powered plants, while 27% supported maintaining existing plants (Iwata 2012a).

One year after the accident, effects on health have been minimal with no radiation-related deaths, even among workers who faced very high exposure vs. 28 workers at Chernobyl who died during the first year (Hayashi et al., 2012). An estimated additional cancer or thyroid problems for 300 to 500 people is predicted at Fukushima (vs. 3,000–6,000 cases of thyroid illnesses at Chernobyl). Of 10,468 people from three towns at high risk—Namie, Iitatate and Kawamata—58% received less than 1 mSv of exposure and 95 % less than 5 mSv. Just 23 people, including 13 nuclear workers, were assumed to have received more than 15 mSv. By comparison, Americans receive, on average, 3 mSv of radiation annually from natural and man-made sources. Of 3,765 Japanese children from the prefecture examined in October and November 2012 for thyroid problems, medical advisers concluded that no one in that group had problems related to radiation exposure (Hayashi et al., 2012).

## 7.3 Terrestrial Resources

From March 16th to March 22nd of 2011, virtually all milk, vegetable and beef samples from Fukushima and environs were above safe limits for human consumption (Wikipedia 2011a).

The radiological dose received by forest biota during the first 30 days of the accident were compared with benchmark values considered safe for ecosystems or wildlife and permitted ecological risk assessment (Garnier-LaPlace et al., 2011). Calculations for forest ecosystems were based on soil samples collected from the zone of greatest atmospheric deposition (Iitate village area 25 to 45 km northwest of the Fukushima site). Soil activity concentrations on March 15th for iodine-131 was 108,000 Bq/kg, with back calculations to the date of deposition estimated to be 430,000 Bq/kg; on March 31st, it was 62,400 Bq/kg for cesium-134 and 72,900 Bq/kg for cesium-137. External irradiation from the contaminated environment, as well as radionuclides incorporated within the organism via root uptake by plants, ingestion and inhalation were summed to produce a combined dose rate—assuming no further atmospheric releases after March 15th. Based on exposure to iodine and cesium radioisotopes, dose rates—in mGy/day—over the first

30 days post-accident were about 1.0 for plants, 1.5 for birds, 2.3 for soil invertebrates and 3.0 for forest rodents. Other radioisotopes were measured in the soils (tellurium-129, tellurium-129m, tellurium-132, cesium-136, iodine-132) and when these isotopes were included in the calculations the total dose rate estimates for the forest biota ranged from 2 to 6 mGy/day or 10 to 100 times greater than dose rates considered safe for terrestrial ecosystems (Garnier-Laplace et al., 2011). On April 11th, it was determined that soil samples contained as much as 400 times the normal level of radiation from communities beyond a 30-km radius from Fukushima (Wikipedia 2011a).

In July of 2011, beef from cattle in Fukushima tainted with radiocesium had reached supermarkets, setting off a public uproar (Tabuchi 2011). The government temporarily banned beef shipments from the region, but the ban was lifted in August 2011 after extensive testing (Tabuchi (2011). In September 2011, elevated levels of radiation in rice crops were detected near the crippled nuclear power plant at Fukushima (Tabuchi 2011). Radioactive substances have already been detected in beef, milk, spinach and tea leaves, leading to recalls and bans on shipments. Rice grown in the city of Nihomatsu, about 67 km from the Fukushima plant contained up to 500 Bq/kg of radiocesium and was considered safe (namely, less than 500 Bq/kg radiocesium) for human consumption. Moreover; rice from more than 400 locations in Fukushima prefecture contained less than 150 Bq/kg radiocesium (Tabuchi 2011).

The Fukushima accident has led to trace amounts of radiation, including iodine-131, cesium 134 and cesium-137 detected in New York State, Alaska, Hawaii, Oregon, California, Montreal and Austria (Wikipedia 2011c). In March, Japanese officials announced that radioactive iodine-131 exceeding safety limits for infants had been detected at 18 water purification plants in Tokyo and five other prefectures. As of July 2011, the Japanese government has been unable to control the spread of radioactive material into the nation's food. Radioactive material has been detected in a range of produce including spinach, tea leaves, milk, fish and beef, up to 320 km (200 miles) from the nuclear plant. Inside the 18 km evacuation zone around the plant, all farming has been abandoned (Wikipedia 2011c).

In July of 2011, local bird populations in 300 locations near Fukushima had dropped by about a third (vs. 2/3 for Chernobyl); the reasons for this event are still unknown (Hayashi et al., 2012).

## 7.4 Aquatic Resources

Severe impacts are likely for the coastal ecosystem adjacent to the Fukushima nuclear plant based on external irradiation from contaminated sediments and activity concentrations reported in seawater (Garnier-Laplace et al., 2011). For example, seawater concentrations of iodine-131 reached 180,000 Bq/L on March 30th and cesium-137 (measured 330 m offshore) activity was 47,000 Bq/L. Activity concentrations decreased rapidly with distance owing to dilution with seawater and over time. Maximum dose rates for iodine-131, cesium-134 and cesium-137 ranged from 210 to 4,600 mGy daily; the lowest being for marine birds and the highest for macroalgae, with intermediate values of 2,600 mGy daily for benthic fish, molluscs and crustaceans. At these high dose rates, marked reproductive effects and death for the most radiosensitive taxa are predicted for all marine wildlife groups living in the near-field contaminant release area. These predictions are based on measured data for only a few radioisotopes among the suite of possible radionuclides that composed the actual aquatic isotopes present including cobalt-58, zirconium-95, molybdenum-99, technetium-99m, ruthenium-105, ruthenium- 106, tellurium-129, tellurium-132, cesium-134, cesium-136, cesium-137, iodine-131, iodine-132, barium-140 and lanthanum-140 (Garnier-Laplace et al., 2011).

Although radioisotope concentrations in fish, shellfish and seaweed could exceed limits for human consumption for weeks, it is unlikely that scientists would be able to detect any genetic effects on marine life as those affected flora and fauna would probably disperse into the Pacific Ocean or die more quickly (Schiermeier 2011). Concentrations of radionuclides in water, sediments, plants and animals need to be measured and the health of the ecosystem monitored. In particular, brown seaweed *Laminaria digitata*, ubiquitous in coastal waters, absorbs iodine 10,000 times greater than the surrounding water and is a good indicator of radioiodine content (Schiermeier 2011).

Based on prior research, it is noteworthy that concentration factors for cesium-137 and strontium-90 were comparatively elevated among freshwater molluscs, crustaceans and teleosts (200 to 4,000) when compared to concentration factors (0.1-23) among marine conspecifics (Table 7.1). Concentration factors for selected transuranics in marine sediments. macroalgae and in fish, were highest in sediments (up to 2,000,000), intermediate in macroalgae (up to 8,000) and lowest in fish (up to 50) (Table 7.2). Similar data are presented for marine algae,

**Table 7.1. Concentration factorsa for cesium-137 and strontium-90 in aquatic organisms (modified from Whicker and Schultz 1982)**

| *Radionuclide and ecosystem* | *Molluscs, whole* | *Crustaceans. whole* | *Fishes, muscle* |
|---|---|---|---|
| **Cesium-137** | | | |
| Freshwater | 600 | 4,000 | 3,000 |
| Marine | 8 | 23 | 15 |
| **Strontium-90** | | | |
| Freshwater | 600 | 200 | 200 |
| Marine | 1 | 4 | 0.1 |

[a] Concentration factor = Bq per gram fresh weight sample/ Bq per mL medium.

**Table 7.2. Approximate maximum concentration factorsa for selected transuranics in marine sediments, macroalgae and fishes (from Morse and Choppin 1991).**

| *Transuranic nuclide* | *Concentration factor* | | |
|---|---|---|---|
| | *Sediments* | *Macroalgae* | *Fish* |
| Neptunium | 1,000 | 5,000 | 10 |
| Plutonium | 100,000 | 2.000 | 40 |
| Americium, Curium, Berkelium, Californium | 2,000,000 | 8,000 | 50 |

[a] Bq per gram fresh weight sample/Bq per mL water.

grazers and predators and 17 elements (Table 7.3), demonstrating the complexity of data interpretation. Data on selected radionuclide transport rates from surface water to lower depths show that the eastern north Pacific Ocean is far lengthier (5.4 to 220.0 years) than either coastal areas (0.5–20.0 years) or upwelling areas (0.3–8.8 years) (Table 7.4).

The irradiation dose rate for the period March to May 2011 was calculated for marine biota in the coastal zone near the destroyed Fukushima reactor and in the open sea (Kryshev and Sazykina 2011). For representative species of fishes, molluscs and seaweeds (species not given) collected in the coastal zone, the irradiation dose rate did not exceed the safe level of 0.4 mGy/h (10 mGy/day), suggesting that marine populations and ecosystems were protected from ionizing radiation. In the open sea, 30 km from the nuclear power plant, these values were orders of magnitude lower (Kryshev and Sazykina 2011).

By June 2011, the levels of radioactive cesium-137 in seawater between 11 and 480 km offshore were safe for aquatic resources and subsequent human consumption (Hayashi et al., 2012). Bottom-

**Table 7.3. Maximum concentration factorsa reported for selected elements in marine organisms at various trophic levels (from Bowen et al., 1971).**

| *Element* | *Algae* | *Grazers* | *Predators* |
|---|---|---|---|
| Silver (Ag) | 1,000 | 20,000 | 3,000 |
| Cadmium (Cd) | 6,000 | 2,000,000 | 10.000 |
| Cerium (Ce) | 4,500 | 300 | 12 |
| Cobalt (Co) | 1,000 | 10,000 | 50,000 |
| Chromium (Cr) | 600 | 300,000 | 3.900 |
| Cesium (Cs) | 50 | 15 | 10 |
| Iron (Fe) | 70,000 | 300,000 | 30,000 |
| Iodine (I) | 7,000 | 70 | 10 |
| Molybdenum (Mo) | 200 | 175 | 200 |
| Manganese (Mn) | 20,000 | 60,000 | 100,000 |
| Nickel (Ni) | 1,000 | 10,000 | 80 |
| Lead (Pb) | 3,000,000 | 2,000,000 | 200,000 |
| Ruthenium (Ru) | 1,000 | 16 | 10 |
| Strontium (Sr) | 90 | 85 | 5 |
| Titanium (Ti) | 0,000 | 20.000 | 3,000 |
| Zinc (Zn) | 3,000 | 100,000 | 20,000 |
| Zirconium (Zr) | 20,000 | 30,000 | 40,000 |

[a] Bq per gram fresh weight tissue/Bq per mL of seawater

**Table 7.4. Time required to transport selected radionuclides added into marine waters at surface from the upper mixed layer by biological transport. Processes include diurnal vertical migration, fecal pellets and sinking of dead mater (modified from Lowman et al., 1971).**

**Time, in years, required to transport radionuclide from surface to bottom**

| *Radionuclide* | *Eastern North Pacific Ocean* | *Coastal areas* | *Upwelling areas* |
|---|---|---|---|
| Manganese-54 | 74.0 | 7.0 | 3.0 |
| Iron 55 and Iron-59 | 7.2 | 0.7 | 0.3 |
| Cobalt-57, Cobalt-58 and Cobalt-60 | 220.0 | 20.0 | 8.8 |
| Zinc-65 | 12.0 | 1.1 | 0.5 |
| Lead-210 | 7.3 | 0.7 | 0.3 |
| Zirconium-95 | 5.4 | 0.5 | 0.2 |

dwelling fish, however, contained elevated cesium levels up to 1,000 times higher than background. Since cesium binds to sediments on the ocean floor, bottom-dwelling fauna are at greater risk, but the consequences are still unresolved (Hayashi et al., 2012).

## 7.5 Fukushima vs. Chernobyl and Three Mile Island

Media coverage of nuclear accidents and radiation releases has changed dramatically over time (Friedman 2011). The internet made an enormous amount of information on Fukushima available far more rapidly than was provided by the Three Mile Island and Chernobyl accidents. Although journalists contributed much of the news on Fukushima, citizens actively participated in blogs and on Facebook, Twitter, YouTube and the like, exchanging views and directing others to important news articles or videos. Consequently, radiation coverage of the Fukushima accident was better than that of Three Mile Island or Chernobyl. Although heavy print and broadcast coverage also followed the Three Mile Island and Chernobyl accidents in 1979 and 1986, respectively, coverage did not grow as quickly or become as vast as the Fukushima event. All of these internet activities, plus traditional print publications and television and radio broadcasts, played a major role in informing people about Fukushima events and related issues such as nuclear energy policies in various countries. The speed of the online dissemination had good and bad points, although information appeared quickly, it was widely distributed without much thought about its accuracy or the credibility of its sources. For example, an early blog asserting that there was no chance that "significant radiation" would be released from the damaged reactor was re-posted on hundreds of websites and message boards and was even used as a link by some reliable media sites (Friedman 2011).

There are currently only two Level 7 nuclear events: Chernobyl and Fukushima. The Japanese government argued that apart from children who contracted thyroid cancer from drinking Chernobyl-contaminated milk, there have been no other health effects (Penney and Selden 2011). However, analysis of the health impact of radioactive land contamination by the Fukushima Daiichi nuclear power plant, as based on official Japanese Ministry of Education, Culture, Sports, Science and Technology data, has shown that over the next 50 years it is possible to expect 400,000 additional cancer patients within a 200-km radius of the plant. Based on the Chernobyl experience, five recommendations applicable to Fukushima were made (as quoted in Penney and Selden 2011). (1) Enlarge the exclusion zone around the plant from 20 km to 50 km. (2) Distribute detailed instructions on effective ways to protect the health of individuals while avoiding additional contamination of food. Monitor all individuals at least once weekly and distribute radioprotectors and decontaminants as needed. (3) Develop recommendations for safe agriculture on the contaminated territories, including milk reprocessing, meat decontamination and biofuel production. (4) Improve existing medical centers and create new ones to deal with the immediate and long-term consequences of irradiation. (5) Organize post-Fukushima life in the contaminated territories (based on Chernobyl) by creating a special powerful interagency state body to oversee problems arising from contamination during the initial complicated years.

Penney and Selden (2011) dispute much of the official data published by the World Health Organization (WHO) and the International Atomic Energy Agency (IAEA 2011) on Chernobyl data. For example, Chernobyl deaths actually numbered 8,930 and another 10,000–25,000 additional fatalities are expected from cancer and leukemia. Among the Chernobyl cleanup crew (termed liquidators and numbering up to 1 million), at least 25,000 have died and 70,000 are disabled. There is no threshold below which ionizing radiation poses no risk and the risk is proportional to the dose (a linear no-threshold). The risk of cancer from radiation is 5% per sievert (as quoted in Penney and Selden 2011), although for children the risk may be 15% (equivalent to 0.13 Sv) and permissible radiation for occupational workers is 0.05 Sv. There are many comparisons between Chernobyl and Fukushima and also many contrasts (as quoted in Penney and Selden 2011). Chernobyl radiation releases were almost entirely to the atmosphere and very little to the ocean; the reverse was true for Fukushima. The Japanese government, by comparison, responded quickly and efficiently in removing contaminated food from the affected areas and evacuating contaminated populations. On the other hand, coverups by TEPCO and recklessness by the Japanese government were noted, as in its decision to expose Fukushima children to 20.0 mSv of radiation on a long-term basis. To date, the Japanese government has failed to respond effectively to critics of policies that pose long-term risks to the nation's children (as quoted in Penney and Selden 2011).

Matson (2011) states that the average total radiation dose to the 114,500 individuals evacuated during the 1986 Chernobyl nuclear accident was 31.0 mSv, that the average exposure of U.S. residents from natural and anthropogenic sources is 6.2 mSv per year, that the estimated total exposure at the boundary of the Three Mile Island site in Pennsylvania during the 1979 accident there was 1.0 mSv or less and that the maximum allowable exposure for U.S. radiation workers is 50.0 mSv/year. The radiation dose at the boundary of the Fukushima Daiichi nuclear plant on March 16, 2011 was 1.9 mSv/hour but ranged up to 400.0 mSv/hour inside the plant. The amount of nuclear fuel in Chernobyl reactor 4 that exploded in 1986 was 190 metric tons releasing 25 to 57 tons of nuclear fuel and fission by-products into the atmosphere and the approximate amount of nuclear fuel in each crippled Fukushima Daiichi reactor is 70 to 100 metric tons (Matson 2011).

The Three Mile Island and Fukushima accidents appear to be dissimilar because they involved different reactor types (Bevelacqua 2012). However, the health physics related lessons learned from Three Mile Island are applicable to Fukushima and can enhance the recovery effort. These actions include fuel damage assessment; facility and equipment assessment; radiological characterization; implementation of an end state recovery plan; improving facility conditions; expanding and redefining the health physics organization; radioactive waste minimization; work control systems; and risk communications (Bevelacqua 2012). The Fukushima cleanup operation is more likely to resemble the protracted cleanup at the Three Mile Island Generating Station in Pennsylvania, where one reactor experienced a partial meltdown in 1979 after valves in its cooling system malfunctioned (Stone 2011). The laborious cleanup, in which about 150 tons of contaminated materials were shipped to the Idaho National Laboratory for storage, took almost 14 years and cost nearly 1 billion dollars (Stone (2011).

Selected fission products in the Chernobyl reactor core and their estimated escape into the environment is well documented (Table 7.5). Similar data for Fukushima are currently unavailable and must be obtained if a reasonable risk assessment to human health is to be attempted.

## LITERATURE CITED

Aarkrog, A. 1990. Environmental radiation and radiation releases. *International Journal of Radiation Biology.* **57:**619–631.

Anzai, K., N. Ban, T. Ozawa and S. Tokonami. 2012. Fukushima Daiichi nuclear power plan accident: Facts, environmental contamination, possible biological effects and countermeasures, *Journal of Clinical Biochemistry and Nutrition.* **50(1):**2–8.

**Table 7.5. Selected fission products in the Chernobyl reactor core and their estimated escape into the environment (modified from Aarkrog 1990; Severa and Bar 1991; and Eisler 1994).**

**Trillions of becquerels (TBq)**

| *Radionuclide* | *In core* | *Escaped*[a] |
|---|---|---|
| Krypton-95 | 33,000 | 33,000 |
| Xenon-133 | 1,700,000 | 1,700,000 |
| Iodine-131 | 1, 300,000 | 260.000 |
| Tellurium-132 | 320,000 | 48,000 |
| Cesium-134 | 190,000 | 19,000–50,000 |
| Cesium-137 | 290,000 | 37,700–100,000 |
| Molybdenum-99 | 4,800,000 | 110,400 |
| Zirconium-95 | 4,400,000 | 140,800 |
| Ruthenium-103 | 4,100,000 | 118,900 |
| Ruthenium-106 | 2,000,000 | 35,000–58,000 |
| Barium-140 | 2,900,000 | 162,400 |
| Cerium-141 | 4,400,000 | 101,200 |
| Cerium-144 | 3,200,000 | 89,600 |
| Strontium-89 | 2,000,000 | 80,000 |
| Strontium-90 | 200,000 | 8,000 |
| Neptunium-239 | 140,000 | 4,200 |
| Plutonium-238 | 1,000 | 30 |
| Plutonium-239 | 850 | 25 |
| Plutonium-240 | 1,200 | 36 |
| Plutonium-241 | 1 ,700,000 | 5,100 |
| Curium-242 | 26,000 | 780 |

[a] Other radionuclides in the Chernobyl escapement include 1,500 TBq of silver-110, 3,000 TBq of antimony-125, 6 TBq of americium-241 and 6TBq of curium-243 and curium-244

Bevelacqua, J.J. 2012. Applicability of health physics lessons learned from the Three Mile Island unit 2 accident to the Fukushima Daiichi accident, *Journal of Environmental Radioactivity.* **105:**6–10.

Bowen, V.T., J.S. Olsen, C.L. Osterberg and J. Ravera. 1971. Ecological interactions of marine radioactivity. pp. 200–222 in U.S. National Academy of Sciences (NAS), *Radioactivity in the Marine Environment,* NAS, Panel on Radioactivity in the Marine Environment, Washington, D.C.

Dalerba, P. 2011. Blood-cell banking for workers at the Fukushima Daiichi nuclear power plant, *The Lancet.* **378(9790):**485.

Eisler, R. 1994. Radiation hazards to fish, wildlife and invertebrates: A synoptic review, *U.S. Department of the Interior, National Biological Service,* Biological Report 26, Contaminant Hazard Reviews Report 29, 124 p. Washington, D.C. 20240.

Fackler, M. 2011a. Japan split on hope for vast radiation cleanup, *New York Times (newspaper),* December 7th, A1 + A12.

Friedman, S.K. 2011. Three-Mile Island, Chernobyl and Fukushima: an analysis of traditional and new media coverage of nuclear accidents and radiation, *Bulletin of the Atomic Scientists.* **67(5)**:55–65.

Garnier-Laplace, J., K. Beaugelin-Seiller and T. G. Hinton. 2011. Fukushima wildlife dose reconstruction signals ecological consequences, *Environmental Science & Technology.* **45 (12)**:5077–5078.

Hayashi, Y., P. Dvorak and R.L. Hotz. 2012. In Japan, relief at radiation's low toll, *Washington Post (newspaper),* March 10th-11th, C 3.

Harlan, C. 2011a. Same town, in a new place, *Washington Post (newspaper),* September 22nd A 10.

Harlan, C. 2011b. Tepco pays price for disaster. *Washington Post (newspaper)* October 10th, A 9.

Harlan. C. 2011d. In stricken Japan, a few bet on revival, *Washington Post (newspaper),* November 7th, A 8.

Hayashi, Y. 2011. Past haunts tally of Japan's nuke crisis, *Wall Street Journal (newspaper),* December 23rd, A1 + A16.

Hayashi, Y. 2012. Displaced Japanese town tries to stay intact, *Wall Street Journal (newspaper),* March 3-4, A 8.

IAEA (International Atomic Energy Agency). 2011. Fukushima nuclear accident update log, iaea.org/...tsunamiupdate01.html

Iwata, M. 2012a. Japan nuclear stress tests fail to assuage public fears, *Wall Street Journal (newspaper).* March 3-4, A 8.

Kryshev, I.I. and T.G. Sazykina. 2011, Evaluation of the irradiation dose rate from marin biota in the region of the destroyed Fukushima reactor (Japan) in March-May 2011, *Atomic Energy.* **111(1)**:41–45 (translated from Russian).

Lowman, F.G., T.R. Rice and F.A. Richards. 1971. Accumulation and redistribution of radionuclides by marine organisms, pp. 161–199 in National Academy of Sciences. Radioactivity in the marine environment. National Academy of Sciences, Panel on radioactivity in the marine environment. Washington, D.C.

Matson, J. 2011. Fast facts about radiation from the Fukushima Daiichi nuclear reactors, *Scientific American,* March 16th.

Morse, J.W. and G.R. Choppin. 1991. The chemistry of transuranic elements in natural waters, *Reviews in Aquatic Science.* **4**:1–22.

Nakanishi, S., J.E. Moore, M. Matsuda, C.E. Goldsmith, W.A. Coulter and J.R. Rao. 2012. Bacterial stress response to environmental radiation relating to the Fukushima radiation discharge event, Japan: Will environmental bacteria alter their antibiotic susceptibility profile? *Ecotoxicology and Environmental Safety.* **7**:169–174.

Niazi, A.K. and S.K. Niazi. 2011. Endocrine effects of Fukushima: Radiation-induced endocrinopathy, *Indian Journal of Endocrinology and Metabolism.* **15(2)**:91–95.

Ohnishi, T. 2012. The disaster at Japan's Fukushima-Daiichi nuclear power plant after the March 11, 2011 earthquake and tsunami and the resulting spread of radioisotope contamination, *Radiation Research.* **177(1)**:1–14.

Osnos, E. 2011. Letter from Fukushima. The fallout. *The New Yorker,* October 17, pp. 46–61.

Penney, M. and M. Selden. 2011. The severity of the Fukushima Daiichi nuclear disaster: Comparing Chernobyl and Fukushima, *Global Research, Asia Pacific Journal,* May 24th.

Schiermeier, Q. 2011. Radiation release will hit marine life, *Nature.* **472**:145–146

Severa, J. and J. Bar. 1991. Handbook of radioactive contamination and decontamination. *Studies in Environmental Science, 47,* Elsevier, New York, 363 p.

Stone, R. 2011. Fukushima cleanup will be drawn out and costly, *Science.* **331(6024)**:1507.

Sun, M., H. Chen, F. Tsai, S. Liu, C. Chen and C.Y. Chen. 2011. Search for novel remedies to augment radiation resistance of inhabitants of Fukushima and Chernobyl disasters: Identifying DNA repair protein XRCC4 inhibitors, *Journal of Biomolecular Structure and Dynamics.* **29(2)**:325–337.

Tabuchi, H. 2011. Radioactivity in Japan rice raises worries. *New York Times* (newspaper), September 25, 10 p.

Tanimoto, T., N. Uchida, Y. Kodama, T. Teshima and S. Taniguchi. Safety of workers at the Fukushima Daiichi nuclear power plant, *The Lancet.* **377(9784)**:968.

Tanoi, K., K.E.N. Hashimoto, K. Sakurai, N. Nihei, Y. Ono and T.M. Nakanishi. 2011. An imaging of radioactivity and determination of Cs-134 and Cs-137 in wheat tissue grown in Fukushima, *Radioisotopes.* **60(8)**:317–322.

Tsumune, D., T. Tsubono, M. Aoyama and K. Hirose. 2011. Distribution of oceanic $^{137}$Cs from the Fukushima Daiichi nuclear power plant simulated numerically by a regional ocean model, *Journal of Environmental Radioactivity,* doi:1016/j.jenvrad.2011.10.007.

Whicker F.W. and V. Schultz. 1982. *Radioecology: Nuclear energy and the environment. Volume I,* CRC Press, Boca Raton, Florida. 228 p.

Wikipedia. 2011. Timeline of the Fukushima Daiichi nuclear disaster, 21 p.

Wikipedia. 2011a. Radiation effects from Fukushima I nuclear accidents. 26 p.

Wikipedia. 2011b. Fukushima Daiichi nuclear power plant, 7 p.

Wikipedia. 2011c. Fukushima Daiichi nuclear disaster, 54 p.

Chapter 8

# Mitigation and Remediation

## 8.1 General

If nuclear power is to grow on the scale required to replace alternate energy sources, major steps are needed to rebuild confidence in users showing that nuclear facilities will be safe from accidents and secure against attacks (Bunn and Heinonen 2011). Operators and regulators world-wide are now reviewing their nuclear safety measures and responding to heightened public concerns. Government's conclusions have ranged from China's plan to continue its massive nuclear effort to Germany's decision to phase out all nuclear energy by 2022. Over the long-term, new reactor designs with greater reliance on safety measures may reduce risks. But for the next few decades, most nuclear energy will be generated by the hundreds of existing reactors and those that will be built with existing designs. The near-term focus is on upgrading safety and security for existing and planned facilities and institutional approaches to find and fix the facilities that pose the highest risks. Six areas are prioritized: (1) higher safety measures; (2) higher security standards; (3) stronger emergency response; (4) strengthened and expanded peer reviews; (5) legally binding requirements; and (6) expanded international cooperation. Every country operating major nuclear facilities should ask for an international team to review its nuclear safety and security arrangements. Treaties governing nuclear safety and security, such as the Convention on the Physical Protection of Nuclear Materials and Facilities, express broad goals but include few specific requirements. Treaties are advised wherein signatories should negotiate specific binding standards for both safety and security, although this is not likely to happen quickly, given the current lack of consensus. There is a clear need for expanded international nuclear safety and security cooperation. The fact that the disaster revealed a range of inadequacies in nuclear safety in Japan, one of the world's wealthiest countries and among those with the longest experience using nuclear energy, highlights the stringent demands for political and institutional stability, regulatory effectiveness and sustained organizational excellence that today's nuclear technologies impose (Bunn and Heinonen 2011).

## 8.2 Japan

It is clear that the Fukushima reactors' abilities to maintain cooling in the event of a prolonged loss of power and to vent dangerous gas buildups were insufficient, as were the operators ability to respond to large scale emergencies and the regulators' degree of independence from the nuclear industry (Bunn and Heinonen 2011). Reactor operators should be required to be better prepared for disaster such as floods and earthquakes, as well as for any events that cause a prolonged loss of electrical power. Regulators should reassess whether design bases reflect the spectrum of plausible disasters requiring safety backfits where necessary and should also require operators to plan responses to events beyond plants' design basis. Reactor operators need to install filtered vents, thus reducing the amount of radiation released if a dangerous pressure buildup in the reactors requires venting of gases. Other safety measures recommended include prevention of spent fuels from melting or burning and proper storing of spent fuels. There is also a need for more stringent standards for protecting nuclear facilities against terrorist sabotage. It is emphasized that a nuclear facility cannot be considered safe, in the sense of posing little risk to humans and the environment unless it also secure. Stronger emergency responses from local police, fire and emergency departments are recommended with regular and realistic exercises to make sure that all key players are aware of their role in a nuclear emergency. Creation of an international emergency response team that is interoperable both domestically and internationally is recommended with management by the nuclear industry (Bunn and Heinonen 2011).

Japan's rural regions have been shrinking since the end of World War II despite successive efforts to revitalize rural society (Matanle 2011). The Fukushima accident may present the Japanese state and society an opportunity to rethink regional revitalization and national energy procurement strategies for a safe, sustainable and compassionate society that Prime Minister Kan set the Reconstruction Design Council (Matanle 2011).

Remedial measures recommended for residential areas include removal of contaminated top soils and resurfacing of roads (Smith 2011). On farms, remediation measures include applying potassium fertilizers to crops to compete with radiocesium uptake and giving 'Prussian blue' (cyanogenic compounds such as ferric ferricyanide) boluses to grazing animals to reduce radiocesium absorption. Remediation has some drawbacks: potentially massive quantities of contaminated waste; huge economic costs; and consumers, for social or psychological reasons, may refuse products grown in contaminated areas even when they meet regulations. The Japanese authorities may have to raise exposure limits, as they have begun to do in allowing doses of 250 mSv for radiation workers and raising public exposures from 1 mSv per year to 5 to 10 mSv per year, levels of radioactivity that millions of people residing in areas of high natural radioactivity routinely endure (Smith 2011).

Removal of Fukushima radiocontaminants from drinking water by various water purifiers and adsorbents was investigated (Sato et al., 2011). Radioactive iodine, cesium, strontium, barium and zirconium are hazardous fission products because of the high yield or relatively long half-life. Iodide, iodate, cesium and barium were removed by all water purifiers at efficiencies of about 85%, 40%, up to 90% and >85%, respectively. These efficiencies lasted for 200 L, which is near the recommenced limit for use of filter cartridges. Strontium was removed by 70 to 100%, but efficiencies of filters decreased with use. Zirconium was removed by only two of the four filters. Synthetic zeolite A4 efficiently removed cesium, strontium and barium, but had no effect on iodine and zirconium. Natural zeolite, mordenite, removed cesium with an efficiency as high as zeolite A4, but removal efficiencies for strontium and barium were far less than those of zeolite A4. Activated carbon had little removal effects on these elements. Authors conclude that water purifiers are recommended for convenient decontamination of drinking water in the home (Sato et al., 2011).

Seven lessons that can be learned from Fukushima are as follows (Schnoor 2011). (1) The worst case scenario is beyond imagination (in Fukushima, earthquake, tsunami, massive power failure; in Diablo Canyon, California, when an earthquake of 8.0 and higher may exceed design capability of the reactor; or a terror event at the Indian Point reactor near New York City). (2) Loss of power, water, or cell phone communication, as is common during earthquake, flood, hurricane, tornado, or terrorist event. (3) Avoid technologies that can melt down or go critical. (4) Don't imagine that it's safe simply because it has been historically safe. Nuclear power has the safest record of any energy source in the U.S. based on the lack of fatalities. However, accidents are frequently common when one is lulled into a sense of security. (5) Don't build reactors if you are unsure how the waste will be disposed. In the U.S., the Yucca Mountain, Nevada, site has been eliminated as a repository for storage of spent fuel rods and fuel reprocessing has been curtailed. (6) Don't assume that spent fuel rods are stored safely. The U.S. notes that hundreds of reactors throughout the world are storing spent fuel rods on-site in vats of water like those at Fukushima. Officials state that this practice is safe because the material is insufficiently enriched to go critical, although it could produce a "dirty bomb" of radiation. (7) And finally, don't finance a project if the government must underwrite the insurance or collateral the investment (Schnoor 2011).

Japan's two-stage test of the stability of its nuclear plants in the event of a natural disaster meets global safety standards, but more work is needed to prepare for the aftermath of a nuclear accident, according to a report released on January 29th by the International Atomic Energy Agency (Obe 2012). The tests are being implemented in two stages. The first assesses reactor responses to natural disasters and the second focuses on their ability to deal with the aftermath of a disaster. Anxious to avoid major power shortages, the Nuclear and Industrial Safety Agency, Japan's main nuclear regulator, is expected to announce its approval of test results at some reactors. Approval would be the first step towards restarting idled reactors once the Japanese cabinet and local communities concur. As of February 2012, 51 of Japan's 54 reactors were offline, eliminating most of the 30 percent of the nation's electricity that comes from nuclear power and raising fears of power shortages in the peak summer period (Obe 2012).

Japan's long and expensive pursuit of a super-efficient fast-breeder nuclear reactor may be abandoned for budgetary reasons (Harlan 2012a). The four decade project has thus far cost more than 13 billion dollars producing only accidents, controversies and a single hour of electricity. In theory, a fast-breeder reactor can run on the reprocessed uranium and plutonium that conventional light-water reactors generate as by-products, thus producing more fuel than they use. Critics of the project maintain that the plutonium fuel and its coolant system of sodium (which explodes on contact with water) pose unnecessary risks. With fast-breeder reactors, Japan could solve its costly resource-scarcity problem and eliminate the need for expensive fuel imports from across the world. Japan's only prototype fast-breeder, a 280 megawatt reactor, is located in Tsuruga, about 400 km west of Tokyo. The reactor is idle and its future uncertain, although not all research has stopped with a goal of creating a fast-breeder by 2050 (Harlan 2012a).

One year after the accident, Japan is foundering (Blustein 2012). Many residents have become increasingly—and irrationally—preoccupied with how radiation from the crippled Fukushima Daiichi power plant might affect them, despite assurances from the government and reputable experts that health risks are minuscule except in areas very close to the plant. Many local governments are refusing to accept for burial millions of tons of rubble left by the tsunami, even if radiation emissions were zero. They are also balking at plans to restart reactors in their jurisdictions. Shelters that once accommodated half a million people are now closed and residents moved into government-provided housing. Reconstruction costs will swell the government's 12 trillion dollar debt, which stood at 212% of gross domestic product (compared with 165% for Greece and 128% for Italy). By 2013, as reconstruction spending mounts, the debt burden in Japan will reach 227 % of gross domestic product, according to the Organization for Economic Cooperation and Development (OECD). A failure to rein in future deficits could lead bond investors to lose confidence in Japan's credit-worthiness, plunging the economy into much worse trouble. In addition, a significantly smaller workforce is anticipated in about 20 years that will be supporting a growing number of seniors. These—and other long-standing problems—need to be addressed if Japan is to rebound from catastrophe (Blustein 2012). One stopgap measure being considered is a doubling of the sales tax which would generate $170 billion yearly (Harlan 2012b).

## 8.3 United States

In a report issued by the U.S. Nuclear Regulatory Commission (NRC) on July 11, 2011, an elite Task Force concluded that a sequence of events such as the Fukushima accident is unlikely to occur in the United States and that some mitigation measures have already been implemented, reducing the likelihood of core damage and radiological releases and that continued operation and continued licensing activities do not pose an imminent risk to public health and safety (Miller et al., 2011). The Task Force stated that voluntary industry initiatives should not serve as substitutes for regulatory requirements but as a mechanism for facilitating and standardizing implementations of such requirements. The Task Force made a dozen specific recommendations in the following five areas: (1) *Clarifying the Regulatory Framework* including the establishment of a logical, systematic and coherent regulatory framework for adequate protection that balances multiple defense systems (defense-in-depth) and risk considerations; (2) *Ensuring Protection* including mandatory reevaluation by licensees and upgrade if necessary for the design-basis seismic and flooding protection of all structures, systems and components of each operating reactor with NRC evaluation of potential enhancements for prevention or mitigation of seismically induced fires and floods; (3) *Enhancing Mitigation* including prevention of power loss; hydrogen control and mitigation inside containment or in other buildings; enhancing spent fuel pool makeup capability and instrumentation for the spent fuel pool; and strengthened onsite emergency response capabilities such as severe accident management guidelines and extensive damage mitigation guidelines. The fourth general area (*Strengthening Emergency Preparedness)* includes recommendations to address prolonged power loss and NRC overview emergency preparedness topics related to decision-making, radiation monitoring and public education. The last general area was (5) *Improving the Efficiency of NRC Programs* a strengthening of NRC regulatory oversight of safety performance of licensee by focusing more attention on defense-in-depth requirements consistent with that framework (Miller et al., 2011).

In the confusion following the earthquake and tsunami on March 11, 2011, the U.S. Nuclear Regulatory Commission (NRC) said that it was standing by to help; however, emails posted on the NRC's web site shows an agency unable to respond and to deal with the American public (Mufson 2012). The emails provide a candid picture of the level of uncertainty and confusion within the U.S. government and indicates that U.S. experts had major divisions about what was going on and how to best mitigate the crisis. While assuring U.S. citizens publicly that there was no danger, the NRC did not disclose one worst case scenario that did not rule out the possibility of radiation exceeding safe levels for thyroid doses in Alaska. In the end, Alaska was not affected. The NRC also found itself in a sensitive spot on the state of pools for spent nuclear fuel, but did not want to share all its background on this subject, much of it classified. By coincidence, the NRC had told the commissioner on March 11th, the day of the accident, that they intended to issue a license extension to the Vermont Yankee nuclear reactor. But the Vermont Yankee reactor was built to the same design as unit 1 at Fukushima Daiichi and the issuance was delayed; however, the NRC publicly approved the 20-year extension of Vermont Yankee's license on March 21st 2011 (Mufson 2012).

U.S. utilities are building only one kind of new reactor, the AP1000, designed by Toshiba's Westinghouse unit (Smith 2012a). It has been chosen by Southern Company and Scana Corporation for projects in Georgia and South Carolina. These reactors use passive safety systems including large reservoirs of water stored above the reactor and spent-fuel pool so that gravity, not pumps, can get water to the reactor core and pool if needed. Existing reactors, by contrast, rely more on active safety systems such as electrically-driven pumps and valves to maintain proper fluid levels and temperatures. If reactors lose electricity, they can run out of coolant in a few days, overheat and even melt down—as was the case for Fukushima. New reactors also incorporate improved materials, such as cement that can better withstand radiation, heat, caustic chemicals and steel alloys that are less susceptible to cracking and corrosion. A major advantage of the new reactors is that they can be built in modular fashion with large sections constructed at factories and assembled at the site. Modular construction is new for the nuclear sector. The 104 existing U.S. reactors in the U.S. were each heavily customized, which made them more expensive to build, harder to inspect and more difficult to maintain. Another advantage is that an operator qualified to work in the control room of one AP1000 should be able to step into the control room of another AP1000 without additional training. By the year 2050 there should be totally passive reactors that will be buried underground so they will pose no threat even if hey have problems. According to the director of the U.S. Department of Energy's Idaho National Laboratory, should a problem arise, the site can be abandoned. In that scenario, the reactor gradually dies down with loss of capital investment; however, environmental hazards are minimal (Smith 2012a).

## LITERATURE CITED

Blustein, P. 2012. A wake-up call Japan ignored, *Washington Post (newspaper)*, March 11th, B 3.

Bunn, M. and O. Heinonen. 2011. Preventing the next Fukushima, *Science*, 333 (6049), 1580-1581.

Harlan, C. 2012a. Japan losing hope of 'dream reactor', *Washington Post (newspaper),* February 1st, A 11.

Harlan, C. 2012b. Japan moves toward doubling of sales tax, *Washington Post (newspaper),* June 27th, A 10.

Matanle, P. 2011. The Great East Japan earthquake, tsunami and nuclear meltdown: Towards the (re)construction of a safe, sustainable and compassionate society in Japan's shrinking regions, *Local Environment.* **16(9)**:823–847.

Miller, C., A. Cubbage, D. Dorman, J. Grobe, G. Holahan and N. Sanfillippo. 2011. Recommendations for enhancing reactor safety in the 21st century. The near-term task force review of insights from the Fukushima Dai-ichi accident, *Available from the U.S. Nuclear Regulatory Commission as document ML 111861807,* 83 p.

Mufson, S. 2012. The NRC's Japan meltdown, *Washington Post (newspaper),* February 7th, A 21.

Obe, M. 2012. Japan girds for summer power cuts, *Wall Street Journal (newspaper),* January 20th, A6.

Sato, I., H. Kudo and S. Tsuda. 2011. Removal efficiency of water purifier and adsorbent for iodine, cesium, strontium, barium and zirconium in drinking water, *The Journal of Toxicological Sciences.* **36(6)**:829–834.

Schnoor, J.L. 2011. Lessons from Fukushima, *Environmental Science & Technology.* **45(9)**:3820.

Smith, J. 2011. A long shadow over Fukushima, *Nature.* **472**:7, doi:10:1038/472007a.

Smith, R. 2012a. Industry alters designs in an effort to make future plants safer, *Wall Street Journal (newspaper),* March 9th, A 10.

CHAPTER 9

# Implications for the Nuclear Reactor Industry

## 9.1 General

The role of carbon emissions in climate change makes clear the necessity for a global reconsideration of energy production (Eisler 2012). Four technologies are currently under consideration: nuclear power, carbon capture and storage, wind power and geoengineering (Poumadere et al., 2011). All of these approaches are often socially controversial and present complex challenges of governance. Support for nuclear energy appears to be conditional upon simultaneous development of other renewable technologies as well as satisfactory disposal of nuclear wastes. The Fukushima accident, for example, greatly increased public concern about the safety and vulnerability of nuclear reactors. Authors conclude that exercises in risk governance should be developed at the national and international levels (Poumadere et al., 2011).

Liquified natural gas (LNG) is now considered a serious alternative to nuclear power world-wide (Iwata and Landers 2012). Japan—which produces less than 4 percent of the natural gas than it consumes—has boosted imports of natural gas by about $8 billion dollars in the final three quarters of 2011. Germany—which has said that it will close its nuclear plants by 2022—opened a new gas line from Russia in November 2011. The BG group in Britain has signed a contract to import natural gas from the United States. Japan was already the world's biggest importer of LNG before the March 2011 earthquake, with most imports from Malaysia, Qatar and Australia. Although LNG is a fossil fuel that contributes to greenhouse gases, Japanese users see LNG as cleaner than coal, less expensive than oil and more readily available than sun or wind power. Canada is exporting LNG from its west coast near Kitimat, British Columbia. Mitsui, a Japanese firm, said that it hopes to sell about 5 million tons a year to Japanese customers of LNG from a field off the coast of Mozambique, or about half the planned output. Global natural gas exploration has increased markedly in recent years (Iwata and Landers 2012).

A year after the Fukushima accident, developing countries with an insatiable thirst for electricity are building nuclear reactors (Dawson et al., 2012). Sixty reactors are currently under construction globally with 163 more on order or planned, according to the World Nuclear Association. While Japan and some European nations prepare to shut down or idle their nuclear plans, the march to build reactors continues in developing countries (Dawson et al., 2012).

## 9.2 Japan

Months after the accident, the nuclear crisis continues (Suzuki 2011). Many technical, social, legal and economic problems need resolution. Major short-term challenges include stabilizing the reactors and managing more than 100,000 tons of contaminated water, as well as cleaning up the site which still contains a large amount of contaminated debris from the accident. Long-term challenges include dealing with spent fuel in the storage pools and damaged fuel in the reactors and finally decommissioning the reactors (Suzuki 2011).

Japanese authorities have admitted that lax standards and poor oversight have contributed to the nuclear accident (Wikipedia 2011c). These authorities have come under fire for their handling of the emergency and for their behavior pattern of withholding damaging information and denying facts about the accident. Authorities apparently wanted to limit the size of costly and disruptive evacuations and to avoid public questioning of the politically powerful nuclear industry. The Japanese public is angered about the alleged campaign to play down the scope of the accident and the potential health risk. The Japanese prime minister—once a proponent of nuclear reactors—took an increasingly anti-nuclear stance in the months following the Fukushima event. In May, he ordered the aging Hamaoka Nuclear

Plant closed over earthquake and tsunami considerations and stated that he would freeze plans to build new reactors and that Japan should reduce and eventually eliminate its dependence on nuclear energy (Wikipedia 2011c). Implications of the nuclear accidents at Fukushima involve questions of nuclear regulatory standards, broader implications on non-carbon emitting energy production, nuclear nonproliferation objectives and community resilience and emergency response against catastrophic events (Hall 2011).

An independent panel appointed by Japan's government assessed the planning, response and aftermath of the Fukushima nuclear disaster; this 500-page report differed sharply from those prepared previously by TEPCO and the Japanese government (Dvorak and Obe 2011a). The report faults the plant's operators and regulators for shoddy planning, bungled responses and poor public communications. Among their findings—based on interviews with more than 450 people—were the following: plant workers didn't have training or manuals to handle severe emergency scenarios; workers failed for hours to notice or report failed safety systems closing cooling water to reactor No. 1; engineers shut down a system supplying cooling water to No 3, leaving the overheating reactor without water for nearly 7 hours; and the regulators' emergency response center never functioned properly (Dvorak and Obe 2011a). Japan said that it will set age limits on nuclear plants to 40 years and require operators to plan for worst-case scenarios (Olsen and Dvorak 2012). One pressing debate is the fate of Japan's many reactors that have been shut down for routine maintenance, then kept idle as distrustful communities have resisted restarting them (Olsen and Dvorak 2012),

Japan reported its first trade deficit since 1980 underscoring the mounting challenges for an economy slowed by natural and nuclear disasters (Harlan 2012; Nakamichi 2012). Export sales were diminished by earthquakes in northern Japan and by the October floods in Thailand, a major base of operations for many Japanese manufacturers. On the import side, Japan was hit by higher energy costs, mainly from an increase in imported oil and gas. The cost of fossil fuels has been rising as demand from emerging economies like China push up prices. With nuclear plants required to shut for maintenance every 13 months, just four of the nation's 54 reactors are still operating in late January 2012. Japanese authorities hope to restart some units in late 2012, although many local governments that host the plants are currently opposed (Harlan 2012; Nakamichi 2012).

Damage to the Fukushima nuclear plant by the recent earthquake and tsunami should stimulate consideration of alternative sources of energy (Sasaki et al., 2011). If managed appropriately, the 25.1 million ha of Japanese forest could be an important source of wood biomass for bioenergy production (Sasaki et al., 2011). Most scientists now disagree with the concept of wood and fossil fuel combustion as this would increase atmospheric levels of $CO_2$ and decrease the amount of $CO_2$ that is taken up by forests—leading to increasing atmospheric temperatures and decreasing ocean pH as the oceans acidify (Eisler 2012). The prime minister announced that Japan should meet its energy needs without nuclear power plants (Takubo 2011) and another powerful Japanese politician claimed that society can function without nuclear energy (Sekiguchi 2012). Despite the seriousness of the Fukushima crisis, Japan has a historical commitment to nuclear power, a fuel cycle that includes reprocessing and breeder reactors and powerful political supporters. Even with a scale-down of nuclear power, the policy of reprocessing spent nuclear fuel will continue as a matter of political inertia (Takubo 2011). In Japan, for example, 82% of the population favored adding more nuclear plants as recently as 2005; today that number is less than 30% (Hoagland 2011).

As of January 20th, 2012, the official overseeing Japan's energy industry said that there may be no nuclear reactors operating when power demands peak this summer, as the country struggles with how far and fast to reduce its reliance on nuclear energy (Obe 2012). Shutdowns for scheduled maintenance have left all but five of Japan's 54 reactors idle, with the stress tests for restarting them behind schedule. Alienating opponents of nuclear power would risk a backlash from businessmen warning of severe economic repercussions unless reactors are restarted by summer (Obe 2012).

Nearly a year after the triple meltdown at the Fukushima Daiichi facility, Japanese decision-makers cannot agree on how to safeguard their reactors against future disasters, or even whether to operate them at all (Harlan 2012a). The nation's system for nuclear decision-making requires the agreement of thousands of officials. Most bureaucrats and politicians in Tokyo want Japan to recommit to nuclear power, but are opposed by reformists and regional governors. The stalemate comes with heavy consequences, especially as reactors are idled, major power companies recorded financial losses and economy-stunting electricity shortages occurred in major manufacturing hubs. These shortages are likely to mount as more reactors are shut down for required maintenance. At the end of January 2012, only four of Japan's 54 reactors were in operation, generating just 8 percent of their potential nuclear power capacity. By the end of April 2012, these last reactors were due to be idled for "stress" testing and Japan—once the world's third-largest nuclear consumer—would be nuclear-free if it is unable to win approval from local communities to restart the idled units. The current debate boils down to how, or whether, the country can guarantee their safety. Japan's nuclear safety agency says that "stress tests", in which computers simulate a reactor's response to earthquakes and tsunamis, will be enough to assess the risks. The Japanese nuclear companies have seen their values drop by as much as 50 percent and are now firing up old thermal plants to produce electricity at higher costs. In recent weeks, officials from the Kensai Electric Power Company (KEPCO)—Japan's largest nuclear operator—with 11 nuclear reactors that supply almost 50 percent of the power to Osaka and Kyoto in the Kensai industrial region has only one functioning reactor today. By the end of summer, a panel of 25 experts plan to draft Japan's new "Basic Energy Plan" that must then be approved by parliament (Harlan 2012a).

Japan's nuclear power crisis, which has cut power-generating capacity by nearly 25 percent, has left the country dependent on aging conventional plants operating well beyond their limits (Iwata 2012; Iwata and Landers 2012). Only three of the 54 reactors are now operating and these may be shut down as early as April 2012. The biggest risk will come when demand peaks during the hot and humid summers. In a typical August, for example, demand rises 20 percent from the May low. In the face of a sudden shortfall, big utilities may shift to thermal power plants fueled by more expensive coal, gas and oil. The Japanese utilities are cautious about any discussion of summer power shortfalls. All nine utilities that have nuclear reactors declined to talk about the possibility of blackouts. One utility plans to keep a 1970s-era oil-fired power unit in service instead of going through with a planned dismantling. Another is working on small-scale portable gas turbines as supplemental power sources (Iwata 2012; Iwata and Landers 2012). However, in February 2012, the Japanese government released $8.9 billion dollars to TEPCO in exchange for the utility's promise to review a rate increase of 17% and to implement more aggressive cost-cutting to deal with the expenses of the Fukushima disaster (Obe 2012c). Japan's economy minister has indicated that the government is prepared to take control of TEPCO, probably by buying a controlling stake in the company and sell off at least some of its power generating facilities. Such a move could boost competition in Japan's energy market, where a group of nine regional utilities have near monopolies over their areas and electricity costs are some of the world's highest (Obe 2012c).

By early March 2012, Japan's biggest banks were finalizing proposals to provide TEPCO with new loans and rollovers worth billions of dollars (Obe 2012d). Under the proposal, TEPCO's main creditors will provide 6.1 billion dollars in new loans, set up 4.9 billion in new credit lines and agree to rollover 2.0 billion dollars. TEPCO, however faces two main problems. It must generate enough profit to pay compensation to those affected by the Fukushima disaster (by raising electricity rates as much as 17 percent) and decommission the ruined plant—undertakings estimated to take years and cost trillions of dollars to complete. Some analysts argue that the utility should be split into two parts; one part that inherits the huge liabilities and compensation costs and a second part made up of the remainder that could tap the debt market for funding (Obe 2012d). In any event, it seems clear that TEPCO and the Japanese nuclear industry are far from defunct.

As of June 2012, there has been little legal fallout from the Fukushima accident, with only 20 lawsuits filed against TEPCO from victims seeking compensation (Harlan 2012b). In Japan, victims and lawyers say the dearth of nuclear-related lawsuits reflects a national mind-set distaste for confrontation and a judicial system that doesn't allow for class-action cases or punitive damages. Instead the vast majority of victims turn directly to TEPCO where the average payout to individuals is $24,000 or they can turn to a government-created mediation center which was established by law after the nuclear accident. Neither route offers victims much leverage Harlan 2012b).

Future scenarios for electricity generation in the year 2030 includes renewable energy, particularly solar and wind power, nuclear power and fossil fuels (Zhang et al., 2012; Obe 2012). Nuclear power accounted for 30 percent of Japan's electricity supply in 2010 and might contribute as much as 40 percent under the Strategic Energy Plan released by the government in 2010 before the Fukushima accident. New projections for 2030 are: 10 to 30 percent renewable energy, up to 40 percent nuclear and fossil fuels the remainder. However $CO_2$ emissions reduction can be realized with an increasing use of nuclear power and renewable energy, with total $CO_2$ emissions ranging from 10 million to 250 million tons in 2030 vs. 290 million tons in 1990 (Zhang et al., 2012). At present, 2 reactors are under construction (Dawson et al., 2012).

In June 2012, Japan ordered a pair of reactors—operated by the Kensai Electric Power Company—back online for the first time since last year's accident amid much chaos and confusion (Obe and Dawson 2012). The decision was based, in part, on the rationale that reactors were necessary to prevent Japan's economy from shrinking as much as 5% by 2030. However, opinion polls consistently showed that more than half the Japanese public were opposed to nuclear power though they fret over anticipated energy shortages and higher electric bills if the reactors remain off. The business community has strongly backed restarts, stating the need for stable power supplies. The deep ambivalence is centered near Ohi, the western town that is home to the first two reactors slated to come back online, with the approval of the town's mayor. A new agency is expected to launch in September is set to draw up new nuclear safety guidelines and could take a stricter line on vetting reactors; until then, current regulators oversee the industry. In the communities surrounding Ohi, only 38% of residents support the restart of the reactors (Obe and Dawson 2012). Seismologists have warned against a nuclear restart at this time because seismic modeling by Japan's nuclear regulator did not properly take into account active fault lines near the Ohi site (Reuters 2012). Renewable energy, such as wind, solar and hydroelectric may account for up to 50 to 60% of Japan's total energy production in 2030, one of the most ambitious targets anywhere (Obe 2012e).

## 9.3 Europe

Risks associated with the peaceful uses of nuclear energy and the safety of nuclear power plants were discussed by Germany and others, within the context of the operating company and the reaction of regulators (Scheuermann et al., 2011). Specifically, the actions taken to reassure people and how believable their statements are and how useful the tools developed after the Chernobyl accident can be applied to assist the decision makers (Scheuermann et al., 2011).

The Fukushima incident spread fears of radiation poisoning around the world, although only one or two of the estimated 25,000 deaths were caused by radiation and the rest by earthquake and tsunami that triggered the nuclear meltdowns. Nevertheless, Germany has ordered its 17 nuclear reactors to shut down by 2022 (Hoagland 2011). Austria, Italy and Switzerland are also considering the phasing out of nuclear power (Osnos 2011; Harlan 2011). The situation is different in the United Kingdom (Wilby et al., 2011). The UK's eight proposed new nuclear power stations are all to be sited on the coast with a total useful life of at least 160 years, amid heightened awareness of inundation risk following the failure of the Fukushima plant. Britain's nuclear developers have factored in the possibility of rising sea temperatures and the more extreme weather events over the next 200 years. Adaptation options for new nuclear and other major long-lived coastal developments are described, despite uncertainty about climate scenarios. Together with routine environmental monitoring, initial flexibility of design and safety margins, sites can be adaptively managed throughout their life cycles (Wilby et al., 2011).

Europe's response to the Fukushima accident differed widely among the member states with the United Kingdom policy makers firm in their decision to increase nuclear power generation in the near future. At the other extreme is Germany. Germany has decided, at least temporarily, to shut down the old generation of nuclear reactors and reexamine the safety of all national nuclear power facilities (Wittneben 2012). The author argues that Germany, in contrast to the UK, faced imminent elections, had stronger media reporting, had increasing trust in renewable technologies, a history of nuclear resistance and a feeling of close cultural proximity to the Japanese (Wittneben 2012).

At present 1 or 2 reactors are under construction in the following countries: France, Finland, Armenia, Bulgaria, Lithuania, Poland and Slovakia (Dawson et al., 2012).

## 9.4 The United States and Canada

The United States currently possesses 104 of the 436 operating reactors worldwide (Hoagland 2011). U.S. reactors pose no imminent threat at present, according to the U.S. Nuclear Regulatory Commission, but about 1/3 of the 104 commercial reactors now operating need to upgrade their safety equipment (Osnos 2011). What had been growing acceptance of nuclear power in the United States was eroded sharply following the Fukushima accident (Wikipedia 2011c).

The still-emerging radiological catastrophe in Japan illuminates the inherent danger and unacceptable risks associated with the continued operations of the Fukushima-style reactors, namely, the General Electric Mark I Boiling Water Reactor (Gunter 2012). There are 32 reactors of the same design operating in the United States today. The author recommends the following: greater opportunities for citizen participation in creating a safe energy policy; the urgent need to permanently close the Mark I fleet; set radiation standards to protect the most vulnerable when establishing radiation exposure standards; re-evaluation and expansion of emergency planning zones around nuclear facilities; and a significant upgrade of the enforcement of safety regulations by the U.S. Nuclear Regulatory Commission (Gunter 2012).

The Great Lakes Basin is a heavily populated region with over 45 aging nuclear reactors (Tilman and Rumiel 2012). Ontario bears the scars of the nuclear industry from the legacy of uranium mining, tailing ponds, uranium refineries, fuel production plants, nuclear power plants and nuclear wastes. Canadian power plants generate over 50% of their electricity, and, as an answer to climate change, Canada is proposing the construction of four additional reactors on Lake Ontario and is refurbishing existing ones to extend their life (Tilman and Rumiel 2012).

Responses to Fukushima from the wealthiest and poorest nations differ sharply, with diminished safety for all (Hoagland 2011). Developed countries that are best equipped to deal with nuclear accidents, notably Europeans and Americans, are abandoning or delaying plans to replace or upgrade electricity-producing nuclear plants and extending the operational life well beyond the original 40-year licensing period. The nuclear industry claims that a Fukushima-type event is unlikely to happen in the United States because few U.S. nuclear power plants are vulnerable to tsunamis (Lyman 2011). To some degree, however, every nuclear plant is vulnerable to natural disaster or deliberate attack and no nuclear plant can withstand an event more severe than the "design-basis accidents" it was engineered to withstand. Most U.S. nuclear plants are subject to greater risks than they were designed to handle, particularly in regard to earthquakes, suggesting that new and existing nuclear plants be upgraded accordingly (Lyman 2011).

At the time of the Fukushima accident, two nuclear reactors were under construction in Texas by NRG, an independent power producer (Kwoh 2012). After Fukushima, the multibillion dollar project was halted and the company shifted focus to its solar and natural gas businesses. NRG will resume work on the reactors dependent on the granting of licenses from the Nuclear Regulatory Commission (Kwoh 2012).

On February 9th, 2012, the U.S. Nuclear Regulatory Commission (NRC) approved a construction and operating license for new nuclear reactors for the first time since 1978, allowing Southern Company to build two new units at its existing site in Vogtle, Georgia (Mufson 2012a). The Vogtle designs met current standards. However, when the NRC provides updated guidelines, all nuclear plants, including Vogtle, would have to comply. The Southern Company expects the project to cost about 14 billion dollars with completion dates in 2016 and 2017 (Mufson 2012a). It will use Toshiba's AP1000 reactors, the design of which the NRC certified in December 2011 (Mufson 2012). The last nuclear reactor built in the United States was the Tennessee Valley Authority's Watts Bar plant, which was started in 1973 and completed in 1996. Today nuclear power provides 20 percent of U.S. electricity, but many plants will reach the end of their license extensions in the next 20 to 30 years. NRC approval could mean rapid approval for other new units, including two that the utility Scana has proposed to build in South Carolina (Mufson 2012). U.S. nuclear construction has been stalled by safety concerns and cost over-runs that followed the Three Mile Island accident in Pennsylvania in 1979 (Smith 2012). Southern's first two reactors at the Vogtle site, 42 km southeast of Augusta, Georgia, were begun in the 1970's and completed in the 1980's. They were supposed to cost $660 million, but totaled $8.87 billion. Southern will have the most reactors at the Georgia site—four—of any plant in the United States when construction is done. Scana Corporation is next in line to receive NRC licenses to build the same type of reactors in South Carolina. But it is questionable whether other units will follow because of capital costs and the low price of natural gas (Smith 2012)..

In March, 2012, comparatively inexpensive natural gas became the energy source of choice (Smith 2012b). Expected additions to U.S. power-generation capacity between 2010 and 2035 by fuel type were: natural gas, 58.1%; wind, 12.8%; coal and other fossil fuels, 7.6%; nuclear 4.3%; hydro power 1.6%; and all other renewable sources 15.6% (Smith 2012b).

Improved carbon capture technology to remove carbon dioxide from coal-fired power plants inexpensively is proceeding with technologies including wet scrubbing, dry scrubbing, gas separation membranes and oxy-combustion, although the systems are still under development (Sweet 2012).

School districts across the country are turning to solar power to cut their electricity costs (Carlton 2012). More than 500 K-12 schools in 48 states have installed solar panels, many of them over the past three years as solar power costs have fallen by more than 33%, according to estimates by the Solar Energy Industries Association, a trade group in Washington, D.C. and GTM Research, a Greentech Media Inc. Unit in Boston. Costs have fallen because of increased production of solar panels, creating a glut of solar panels. The generating capacity of new installations has risen from 400 megawatts in 2008 to about

1,900 megawatts in 2011, according to the solar trade group and GTM Research (Carlton 2012).

In early 2012, China's ENN Group said it intends to build a $5 billion ecological center in Nevada that would include a solar-panel factory (Ball 2012). Skeptics question the viability of the project, given the glut in solar-panel manufacturing capacity. But other state-owned Chinese energy companies also indicated that they are shopping for U.S. clean-energy technologies and projects. And the Obama administration is pushing for more Chinese investment in the U.S. clean-energy industry (Ball 2012).

## 9.5 All Others

One year after Fukushima, 26 new reactors were under active construction in China, 10 in Russia, 7 in India, 3 in South Korea, 2 in Taiwan, 2 in Pakistan and 1 in Argentina, with many more planned (Dawson et al., 2012).

The share of nuclear power to the total commercial primary energy in India was about 1.5% in 2004 and this is expected to increase to 5.9% in 2032 (Chokshi 2011). But some Indian government officials believe that renewable energy—such as solar—is likely to be cheaper than nuclear power, especially if all the subsidies, including insurance are taken into account, i.e., the Indian nuclear liability law limits damage to $330 million vs. an estimated $130 billion for the Fukushima accident. Other factors under consideration include potential evacuation of millions (vs. 200,000 at Fukushima) and the inordinate strain on their medical facilities (Chokshi 2011).

In general, however, the global nuclear industry seems unperturbed by the Fukushima Daiichi accident (Peltier 2011). Ten new nuclear plants went online over the past two years and at least 90 new nuclear plants are scheduled to enter service by 2030 (Peltier 2011). However, many countries have advised their nationals to leave Tokyo, citing the risk associated with the nuclear plants' ongoing accident (Wikipedia 2011c). International experts have said that a work force numbering in the thousands may take decades to clean up the area. Stock prices have dropped of many energy companies reliant on nuclear sources, while stock prices of renewable energy companies have increased. (Wikipedia 2011c).

Brazil is now reevaluating decisions to expand their nuclear parks including greater emphasis on other forms of energy (Goldemberg

2011).The pro-nuclear sentiment that has escalated in recent years as concern mounted in developed countries about atmospheric pollution caused by carbon dioxide from combustion of fossil fuels (Eisler 2012) has largely dissipated as a result of Fukushima.

Nevertheless, Japan's nuclear technology companies are increasingly looking to overseas markets, hopeful that foreign governments still trust Japanese technology and safety claims in the aftermath of Fukushima (Harlan 2011). Negotiations are ongoing with Vietnam, Jordan and Lithuania to build nuclear reactors for these and other power-starved nations. In questioning its own reliance on nuclear power (54 seaside reactors), while endorsing its nuclear export efforts, Japan finds itself pulled by contradictory claims. At this time, however, Asian and Middle Eastern countries still have plans for major nuclear reactors and China, alone, has about 25 reactors under construction (Harlan 2011). Developing countries like China and India with little nuclear experience are adding about 80 new reactors in the next 20 years. The main threat of nuclear damage to the planet no longer comes from Russia and the United States, but from nuclear bureaucracies in the Middle East and the Asian sub-continent, as well as terror networks intent on acquiring nuclear weapons (Hoagland 2011). In short, the proliferation of nuclear reactors across Asia and elsewhere, is certain to facilitate nuclear weapons proliferation as well.

Fukushima should contain lessons not just for Japan but for all 31 countries using nuclear power (Suzuki 2011).

## LITERATURE CITED

Ball, J. 2012. Beneath a war of words, money paints a different China-U.S. picture, *Wall Street Journal (newspaper)*, June 18th, R 5.

Carlton, J. 2012. The enlightened classroom, *Wall Street Journal (newspaper)*, June 18th, R 3.

Chokshi, A.H. 2011, Fukushima: India at crossroads, *Current Science.* **100(11):**1603.

Dawson, C., B. Spaegele and S. Williams. 2012. Nuclear pushes on despite Fukushima, *Wall Street Journal (newspaper)*, March 9th, A 1, A 10.

Dvorak, P. and M. Obe. 2011a. Panel finds serious errors in Japan, *Wall Street Journal (newspaper)*, December 27th, A7.

Eisler, R. 2012. *Oceanic acidification. A comprehensive overview.* Science Publishers, Enfield, New Hampshire, 252 p.

Goldemberg, J. 2011. Perspectives for nuclear energy in Brazil after Fukushima, *Brazilian Journal of Physics.* **41:**103–106.

Gunter, P. 2012. "Freeze our Fukushimas"; Why we must permanently close the GE Mark I boiling water reactors, Abstract in The Lessons of Fukushima: A Symposium for Education, Collaboration, Inspiration, February 2, 2012, Willamette University of Law, Salem, Oregon, (http://www.willamette.edu/events/ fukushima/).

Hall, H.L. 2011, Fukushima Daiichi: implications for carbon-free energy, nuclear nonproliferation and community resilience, *Integrated Environmental Assessment and Management.* **7**:406–408, doi:10.1002/ieam.225.

Harlan, C. 2011. Japan's contradiction on nuclear power, *Washington Post (newspaper),* November 17th, A 8.

Harlan, C. 2012. Japan reports first trade deficit in over 3 decades, *Washington Post (newspaper),* January 25th, A 15.

Harlan, C. 2012a. In Japan, officials can't agree on future of nuclear power, *Washington Post (newspaper),* January 26th, A 12.

Harlan, C. 2012b. Little legal fallout from Japan's nuclear crisis, *Washington Post (newspaper),* June 25th, A 8.

Hoagland, J. 2011.Dangerous fallout. The overreaction to Fukushima has focused attention on the wrong issue. *Washington Post* (newspaper), October 7th, A20.

Iwata, M. 2012. Japan nuclear crisis pushes plants to limits, *Wall Street Journal (newspaper),* February 12th, A 12.

Iwata, M and P. Landers. 2012. Crisis in Japan transforms natural-gas market, *Wall Street Journal (newspaper),* February 12th, A 1, A 12.

Kwoh, L 2012. Chief leads NRG from nuclear to solar, gas, *Wall Street Journal (newspaper),* March 7th, B 7.

Lyman, E.S. 2011. Surviving the one-two nuclear punch: Assessing risk and policy in a post-Fukushima world, *Bulletin of the Atomic Scientists.* **67(5)**:47–54.

Mufson, S. 2012a. NRC approves 2 new reactors in Ga., *Washington Post (newspaper),* February 10th. A 14.

Nakamichi, T. 2012. Japan post first trade deficit since '80, *Wall Street Journal (newspaper),* January 22nd, A 13.

Obe, M. 2012. Japan girds for summer power cuts, *Wall Street Journal (newspaper),* January 20th, A6.

Obe, M. 2012c. Tokyo ends Tepco funding standoff, *Wall Street Journal, (newspaper),* February 14th, B 4.

Obe, M. 2012d. Japanese utility Tepco to get help from banks, *Wall Street Journal (newspaper),* March 7th, B 5.

Obe, M. 2012e. First the iPhone. Now renewables, *Wall Street Journal (newspaper),* June 18th, R 7.

Obe, M. and C. Dawson. 2012. Nuclear-restart plans divide Japan, *Wall Street Journal (newspaper),* June 18th, A5.

Olsen, K. and P Dvorak. 2012. Japan plans age limits, tougher tests for nuclear plants, *Wall Street Journal (newspaper),* January 7-8, A9.

Osnos, E. 2011. Letter from Fukushima. The fallout. *The New Yorker*, October 17, pp. 46–61.

Peltier, R. 2011. Four plants demonstrate global growth of nuclear industry, *Power.* **155(11)**:32.

Poumadere, M., R. Bertoldo and J. Samadi. 2011. Public perceptions and governance of controversial technologies to tackle climate change: Nuclear power, carbon capture and storage, wind and geoengineering, *Wiley Interdisciplinary reviews—Climate Change.* **2(5)**:712–727.

Reuters. 2012. Seismologists warn Japan against nuclear restart, http://news.yahoo.com/seismologists-warn-japan-against-nuclear-restart-104131025.html

Sasaki, N., T. Owari and F.E. Putz. 2011. Time to substitute wood bioenergy for nuclear power in Japan, *Energies.* **7**:1051–1057.

Scheuermann, W., A. Piater, C. Krass, A. Lurk, T. Wilbois and Y. Ren. 2011. Modeling consequences of the accident at Fukushima, *International Journal of Nuclear Power.* **56**:325–331.

Sekiguchi, T. 2012. Japanese ex-premier is nuclear activist, *Wall Street Journal (newspaper)*, January 28th, A 9.

Smith, R. 2012. New risks for nuclear plants, *Wall Street Journal (newspaper)*, February 1st, A.3.

Smith, R. 2012b. Cheap natural gas unplugs U.S. nuclear-power revival, *Wall Street Journal (newspaper)*, March 16th, A1, A10.

Suzuki, T. 2011. Deconstructing the zero-risk mindset: The lessons and future responsibilities for a post-Fukushima nuclear Japan. *Bulletin of the Atomic Scientists.* **67(5)**:9–18.

Sweet, C. 2012. Carbon capture, the next generation, *(Wall Street Journal (newspaper)*, June 18th, R 9.

Takubo, M. 2011. Nuclear or not? The complex and uncertain politics of Japan's post-Fukushima energy policy, *Bulletin of the Atomic Scientists.* **67(5)**:19–26.

Tilman, A. and L. Rumiel. 2012. From Fukushima to the Great Lakes Basin. Abstract *in* The Lessons of Fukushima: A Symposium for Education, Collaboration, Inspiration, February 24-25, 2012, Willamette University, College of Law, Salem, Oregon, (http://www.willamette.edu/events/fukushima/).

Wikipedia.. 2011c. Fukushima Daiichi nuclear disaster, 54 p.

Wilby, R.L., R.J. Nicholls, R. Warren, H.S. Wheater, D. Clarke and R.J. Dawson. 2011. Keeping nuclear and other coastal sites safe from climate change, *Proceedings of the Institution of Civil Engineers—Civil Engineering,* **164(3)**:129–136.

Wittneben, B.B.F. 2012, The impact of the Fukushima nuclear accident on European energy policy, *Environmental Science & Policy.* **15**:1–3.

Zhang, Q., K.N, Ishihara, B.C. Mclellan and T. Tezuka. 2012. Scenario analysis on future electricity supply and demand in Japan, *Energy.* **38**:376–385.

CHAPTER 10

# Concluding Remarks

The March 11th, 2011, earthquake-tsunami-nuclear disaster that struck Japan killed thousands, caused enormous property damage and left as many as 300,000 homeless. Japanese governmental authorities and the Japanese general population faced the emergency with remarkable civility, compassion and courage.

Earthquake prediction and prevention remain inexact sciences despite the billions of dollars expended on these topics by Japan alone. Future emphasis should continue to be devoted to prediction and also to prevention, specifically, building regulations for earthquake-resistant materials and construction designs.

Major tsunamis are both infrequent and destructive. More research seems merited on their cause and prevention. The Fukushima tsunami was especially destructive in terms of lives lost, property damage and homelessness. The reactors were inundated by the 15-m wave that flooded the diesel generators, eventually leading to nuclear meltdown and significant loss of radioactivity to the environment. Damage could have been lessened with higher retaining walls (more than 15-m), with generators initially removed to higher ground or inland and to better-trained reactor operators. It is clear that future reactor operators will need more rigorous training and experience to cope with tsunami-like emergencies.

The Fukushima nuclear facility should be decommissioned and the remaining reactors—in Japan and elsewhere—retrofitted to conform to internationally-recognized safety standards. The Tokyo

Electric Power Company may face a large fine for alleged negligence and managerial indifference and could be responsible for expenses of cleanup operations and costs of resettling evacuees. This financial burden may result in the bankruptcy of TEPCO.

Since radiation releases are expected to continue—albeit at a reduced rate—continued monitoring of human evacuees is strongly recommended. Monitoring is also warranted of terrestrial food crops, marine products of commerce and air, water and wildlife. Japanese authorities are to be commended for promptly removing radio-contaminated food from various outlets that were destined for human consumption

Radiological criteria for the protection of human health are currently predicated on the linear-no-threshold (LNT) hypothesis based on the assumption that exposure to any amount of ionizing radiation would be reflected in an increase in cancer frequency. The LNT hypothesis needs to be reexamined and possibly lowered, in view of the recent findings that low doses of radiation administered at low rates may actually be radioprotective. Additional and more intensive research is recommended on potential radioprotective chemicals.

On a global scale, the nuclear industry seems undeterred by Fukushima. Major reactors of advanced design and safety are under construction or in planning in many parts of the world, including the United States. Nuclear fuel will probably continue to be a significant and growing source of energy for most nations. In Japan, however, nuclear will probably remain a significant, but reduced, source of power with an increasing use of alternative energy sources such as wind, solar, hydro, fossil and geothermal.

Finally, suitable technology should be developed to effectively and efficiently remove contaminated soils and sediments, thus enabling evacuees to return to their villages decades before radioisotopes naturally decay to safe levels.

This book documents events only during the first 15 months following the March 11th, 2011 tragedy and is therefor quite limited in scope. Fukushima remains an ongoing event that will, without doubt, result in the publication of many books and more importantly, result in major improvements in reactor design and safety to prevent reoccurrence on any scale.

# Glossary

*(modified from Eisler 1994, 2007[a] and Wikipedia 2011[b])*

**Actinides.** Elements of atomic numbers 89 to 103: actinium, thorium, protactinium, uranium, neptunium, americium, curium, berkelium, californium, einsteinium, fermium, mendelevium, nobelium, lawrencium (Ac, Th, Pa, U, Np, Am, Cm, Bk, Cf, Es, Fm, Md, No, Lw).

**Activity**. The activity of a radioactive material is the number of nuclear disintegrations per unit time. Up to 1977, the accepted unit of activity was the curie (Ci), equivalent to 37 billion disintegrations/second (d/s)—a number that approximated the activity of 1 gram of radium-226. The present unit of activity is the becquerel (Bq) equivalent to 1 d/s.

**Alpha particles** An alpha particle is composed of 2 protons and 2 neutrons and a charge of +2; essentially it is a helium nucleus without orbital electrons. Alpha particles usually originate from the decay of radionuclides of atomic number >82 and are detected in samples that contain uranium, thorium, or radium. Alpha particles react strongly with matter and consequently produce large numbers of ions per unit length of their paths. As a result, they are not very penetrating and traverse only a few cm of air. Alpha particles are unable to penetrate clothing or the outer layer of skin; however, when internally deposited alpha particles are often more damaging than most other types of radiations because comparatively large amounts of energy are transferred in a small volume of tissue. Alpha particle absorption involves ionization and orbital electron excitation, Ionization occurs whenever the particle is sufficiently close to an electron to pull it from its orbit. The alpha also loses kinetic energy by exciting orbital electrons with interactions that are insufficient to cause ionization.

**Atom**. The smallest part of an element that has all the properties of that element. An atom consists of one or more protons and neutrons in the nucleus and of one or more electrons.

**Atomic number**. The number of electrons outside the nucleus of a neutral (nonionized) atom plus the number of protons in the nucleus.

**Becquerel (Bq).** The present accepted unit of activity is the becquerel equivalent to one disintegration/second (1 dps). About 0.37 Bq=1 picocurie.

**Beta particles.** Beta particles are electrons that are spontaneously ejected from the nuclei of radioactive atoms during the decay process. They may be either positively or negatively charged. A positively charged beta, called a *positron*, is less frequently encountered than its negative counterpart, the *negatron*. The *neutrino*, a small particle, accompanies beta emission. The neutrino has little mass and is electrically neutral. But neutrinos conduct a variable part of the energy of transformation and account for the variability in kinetic energies of beta particles that are emitted from a given radionuclide. Positrons are emitted by many of the naturally and artificially produced radionuclides; they are considerably more penetrating than alpha particles but less penetrating than X-rays and gamma rays. Beta particles interact with other electrons and with nuclei in the travel medium. The ultimate fate of a beta particle depends on its charge. Negatrons, after their kinetic energy is spent, combine with a positively charged ion or become free electrons. Positrons also dissipate kinetic energy through ionization and excitation; the collision of positrons and electrons causes annihilation and release of energy that is equal to the sums of their particle masses.

**Breeder reactor**. A nuclear chain reactor in which transmutation produces a greater number of fissionable atoms than the number of consumed parent atoms. Breeders were at fist considered superior to light water reactors because of their superior fuel economy. Interest waned after the 1960s as more uranium reserves were found and new methods of uranium enrichment reduced fuel costs.

**Cosmic rays.** Highly penetrating radiations that originate in outer space.

**Curie (Ci).** The Ci is equal to that quantity of radioactive material producing 37 billion nuclear transformations per second. One millicurie (mCi)= 0.001 Ci; 1 microcurie (*u*Ci) = 1 millionth of a Ci; 1 picocurie (pCi) = 0.037 disintegrations/s. About 27 pCi = 1 becquerel (Bq).

**Decay.** Diminution of a radioactive substance because of nuclear emission of alpha or beta particles or of gamma rays.

**Decay product.** A nuclide resulting from the radioactive disintegration of a radionuclide and found as a result of successive transformations in a radioactive series. A decay product may be either radioactive or stable.

**Effective dose equivalent.** The weighted sum, in sieverts, of the radiation dose equivalents in the most radiosensitive organs and tissues, including gonads, active bone marrow, bone surface cells and the lung.

**Electron.** An electron is a negatively charged particle with a diameter of $10^{-12}$ cm. Every atom consists of one nucleus and one or more electrons. Cathode rays and negatrons are electrons.

**Fission.** The splitting of an atomic nucleus into two fragments that normally releases neutrons and gamma rays. Fission may occur spontaneously or may be induced by capture of bombarding particles. Primary fission products usually decay by beta particle emission to radioactive daughter products. The chain reaction that may result in controlled burning of nuclear fuel or in an uncontrolled nuclear weapons explosion results in the release of 2 or 3 neutrons per fission. Neutrons cause additional fissile nuclei in the vicinity to fission, which produces still more neutrons that in turn produce still more fissions. The speed of the chain reaction is governed by the density and geometry of fissile nuclei and of materials that slow or capture the neutrons. In nuclear reactors, neutron-absorbing rods are inserted to various depths into the reactor core. Theoretically, a nuclear explosion is not physically possible in a reactor because of fuel density, geometry and other factors.

**Fusion**. A nuclear reaction in which smaller atomic nuclei or particles combine to form larger nuclei or particles with release of energy from mass transformation.

**Gamma rays**. Gamma rays have electromagnetic wave energy that is similar to but higher than the energy of X-rays, Gamma rays are highly penetrating and able to traverse several centimeters of lead. See **Photons**.

**Genetically significant dose.** A radiation dose that, if received by every member of the population, would produce the same total genetic injury to the population as the actual doses that are received by the various individuals

**Grey (Gy).** 1 Gy = 1 J/kg = 100 rad.

**Half-life.** The average time in which half the atoms in a sample of radioactive element decay,

**Ion.** An atomic particle, atom, or chemical radical with an either negative or positive electric charge.

**Ionization.** The process by which neutral atoms become either positively or negatively electrically charged by the loss or by the gain of electrons.

**Isomer**. One of two or more radionuclides with the same mass number and the same atomic number but with different energies and radioactive properties for measurable durations.

**Isotope.** One of several radionuclides of the same element (i..e., with the same number of protons in their nuclei) with different numbers of neutrons and different energy contents. A single element may have many isotopes. Uranium, for example may appear naturally as uranium-234 (142 neutrons), uranium-235 (143 neutrons), or uranium 238 (148 neutrons); however, each uranium isotope has 92 protons.

**Joule (J)**. 1 J = $10^7$ ergs

**Light water reactor LWR).** A type of thermal reactor that uses water as its coolant and a neutron moderator, Thermal reactors are the most common type of nuclear reactors and LWR reactors are the most common type of thermal reactor. There are three varieties of LWR: the pressurized water reactor (PWR), the boiling water reactor (BWR) and the supercritical (SCWR) water reactor

**Linear hypothesis.** The assumption that any radiation causes biological damage in the direct proportion of dose to effect.

**Mass number.** The total number of neutrons and protons in the nucleus of the element, which is equal to the sum of the atomic number and the number of neutrons.

**Neutron.** Neutrons are electrically neutral particles that consist of an electron and a proton are not affected by the electrostatic forces of the atom's nucleus or orbital electrons, Because they have no charge, neutrons readily penetrate the atom and may cause a nuclear transformation, Neutrons are produced in the atmosphere by cosmic ray interactions and combine with nitrogen and other gases to form carbon-14, tritium and other radionuclides. A free neutron has a half-life of about 19 minutes, after which it spontaneously decays to a proton, a beta particle and a neutrino. A high energy neutron that encounters biological material is apt to collide with a proton with sufficient force to dislodge the proton from the molecule. The recoil proton may then have sufficient energy to cause secondary damage through ionization and excitation of atoms and molecules along its path.

**Nucleus.** The dense central core of an atom in which most of the mass and all of the positive charge is concentrated. The charge on the nucleus distinguishes one element from another.

**Photons.** X-rays and gamma rays, collectively termed photons, are electromagnetic waves with shorter wavelength than other members of the electromagnetic spectrum such as visible radiation, infrared radiation and radio waves. X- and gamma photons have identical properties, behavior and effects. Gamma rays originate from atomic nuclei, but X-rays arise from the electron shells. All photons travel at the speed of light, but energy is inversely proportional to wavelength. The energy of a photon directly influences its ability to penetrate matter. Many types of nuclear transformations are accompanied by gamma-ray emission. For example, alpha and beta decay of many radionuclides is frequently accompanied by gamma photons. When a parent radionuclide decays to a daughter nuclide, the nucleus of the daughter frequently contains excess energy and is unstable; stability is usually achieved through the release of one or more gamma photons, a process called isometric transition. The daughter nucleus decays from one energy state to another without a change in atomic number or weight. The most probable fate of a photon with an energy higher than the energy of an encountered electron is photoelectric absorption in which the photon transfers its energy to the electron and photon existence ends. As with ionization from any process, secondary radiations that are initiated by the photoelectron produce additional excitation or orbital electrons.

**Positron**. A positively charged particle of equal mass to an electron. Positrons are created either by the radioactive decay of unstable nuclei or by collision with photons.

**Proton.** A positively charged subatomic particle with a mass of 1.67252 x $10^{-24}$ that is slightly less than the mass of a neutron but about 1.836 times greater than the mass of an electron, Protons are identical to hydrogen nuclei; their charge and mass make them potent ionizers.

**Radiation.** The emission and propagation of energy through space or through a material medium in the form of waves. The term also includes subatomic particles such as alpha, beta and cosmic rays and electromagnetic radiation.

**Radiation absorbed dose (rad).** Radiation-induced damage to biological tissue results from the absorption of energy in or around the tissue. The amount of energy absorbed in a given volume of tissue is related to the types and numbers of radiations and the interactions between radiations and tissue atoms or molecules. The fundamental unit of the radiation absorbed dose is the rad; 1 rad = 100 erg (absorbed)/g material. In the latest nomenclature 100 rad = 1 grey (Gy).

**Radiation dose.** The term radiation dose can mean several things, including absorbed dose, dose equivalent, or effective dose equivalent. The absorbed dose of radiation is the imparted energy per unit mass of the irradiated material. Until 1977, the *rad* was the unit of absorbed dose wherein 1 rad = 0.01 joule/kg. The present unit of absorbed dose is the grey (Gy), equivalent to 1 joule/kg. Thus, 1 rad = 0.01 joule/kg = 0.01Gy. Different types of radiation have different Relative Biological Effectiveness (RBE). The RBE of one type of radiation in relation to a reference type of radiation (usually X or gamma) is the inverse ratio of the absorbed doses of the two radiations needed to cause the same degree of the biological effect for which the RBE is given. Regulatory agencies have recommended certain values of RBE for radiation protection and absorbed doses of various irradiations are multiplied by these values to arrive at radio protective doses. The unit of this weighted absorbed dose is the roentgen equivalent man (rem). The dose equivalent is the product of the absorbed dose and a quality factor (Q) and its unit is the rem. The quality factor is a function of the capacity to produce ionization, expressed as the linear energy transfer (LET). A Q value us assigned to each type of radiation: 1 to X-rays, gamma rays and beta particles; 10 to fast neutrons; and 20 to alpha particles and heavy particles. The new unit of the effective dose equivalent is the sievert (Sv), replacing rem, In addition to absorbed dose and dose equivalent, there is also the exposure. Exposure is total electrical charge of ions of one sign produced in air as electrons liberated by X or gamma rays per unit mass of irradiated air, the unit of exposure is Coulomb/

kg, but the old unit, the roentgen (R) is still in use. One roentgen = 2.58 x $10^{-4}$ Coulomb/kg.

**Radioactivity**. The process of spontaneous disintegration by a parent radionuclide, which forms a daughter nuclide. When half the radioactivity remains, that time interval is designated the half-life (Tb ½) The Tb ½ value gives some insight into the behavior of a radionuclide and into its potential hazards.

**Radionuclide.** An atom that is distinguished by its nucleus composition (number of protons, number of neutrons, energy content), atomic number, mass number and atomic mass

**Relative biological effectiveness (RBE).** The biological effectiveness of any type of ionizing radiation in producing a specific damage i.e., leukemia anemia, carcinogenicity (see **Radiation dose**)

**Rem. Roentgen equivalent man.** The amount of ionizing radiation of any type that produces the same damage to humans as 1 roentgen of radiation. In the latest nomenclature 100 rem = 1 Sievert (Sv). One rem = 1 roentgen equivalent physical (rep/relative biological effectiveness (RBE), In the latest nomenclature, 100 rem = 1 Sievert (Sv).

**Richter scale.** A numerical scale for expressing the magnitude of an earthquake on the basis of seismograph oscillations. The more destructive earthquakes typically have magnitudes between 5.5 and 8.9; the scale is logarithmic and a difference of one represents an approximate thirtyfold difference in magnitude.

**Roentgen equivalent physical (rep)**. One rep is equivalent to the amount of ionizing radiation of any type that results in the absorption of 93 ergs/g and is approximately equal to 1 roentgen of X-radiation in soft tissue.

**Sievert (Sv).** New unit of dose equivalent. 1 Sv=100 rem = 1 J/kg. See Radiation dose. 1 millisievert (1 mSV) = 1/1,000th of a Sievert. 1 microsievert (1 *u*Sv) = 1/ millionth of a Sievert.

**Specific activity.** The ratio between activity and the mass of material giving rise to the activity. Biological hazards of radionuclides are directly related to their specific activity and are expressed in Bq/kg mass.

**Threshold hypothesis**. A radiation-dose-consequence hypothesis that holds that biological radiation effects occur only above some minimum dose.

**Transmutation.** A nuclear change that produces a new element from an old one.

**Transuranic elements.** Elements of atomic number>92. All are radioactive and produced artificially; all are members of the actinide group

**X-rays.** See **Photons.**

---

a Eisler, R. 1994. Radiation hazards to fish, wildlife and invertebrates: a synoptic review, *U.S. Department of the Interior, National Biological Service, Biological Report 26, Contaminant Hazard Reviews Report 29,* 1–124.

Eisler, R. 2007. Chapter 27 In Eisler's Encyclopedia of Environmentally Hazardous Chemicals, *Elsevier, Amsterdam,* 677–736.

b ben.wikipedia.org/wiki

# Index

## B



## C

## F



## G

## O



## P



## Q



## R

U



V



W

**X**



**Y**



**Z**

# Color Plate Section

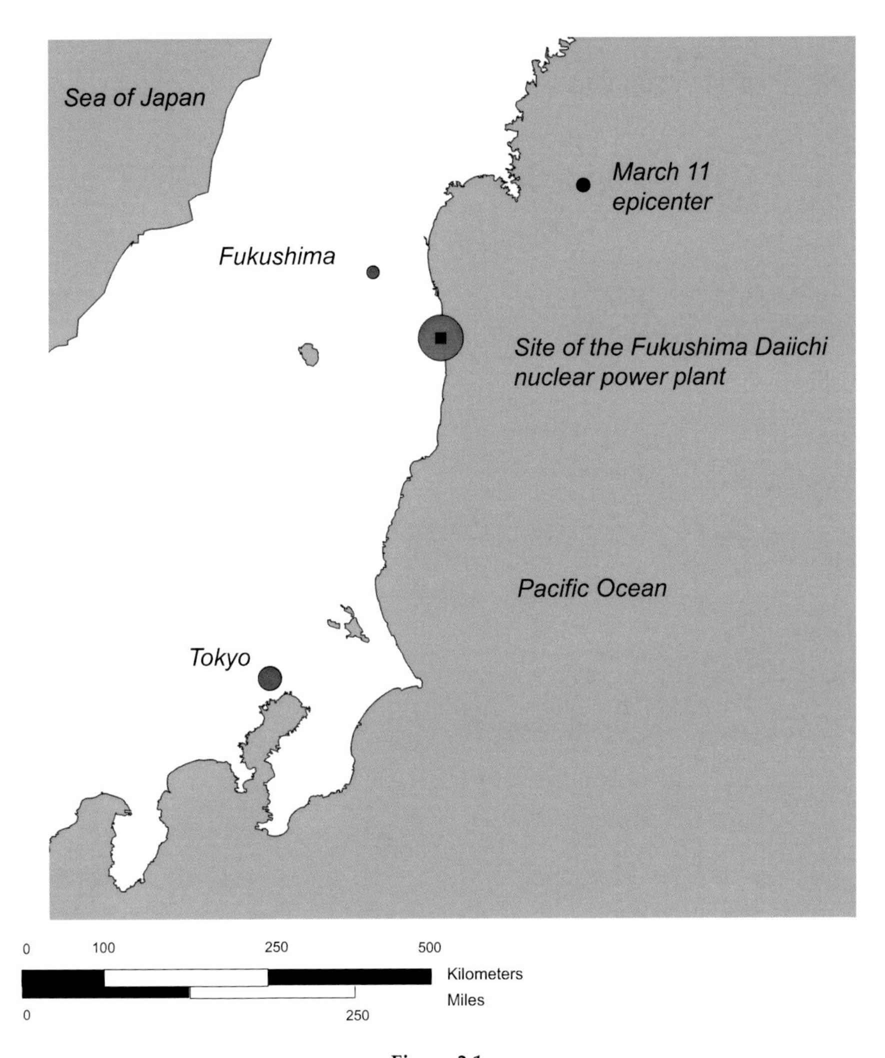


Figure 2.1

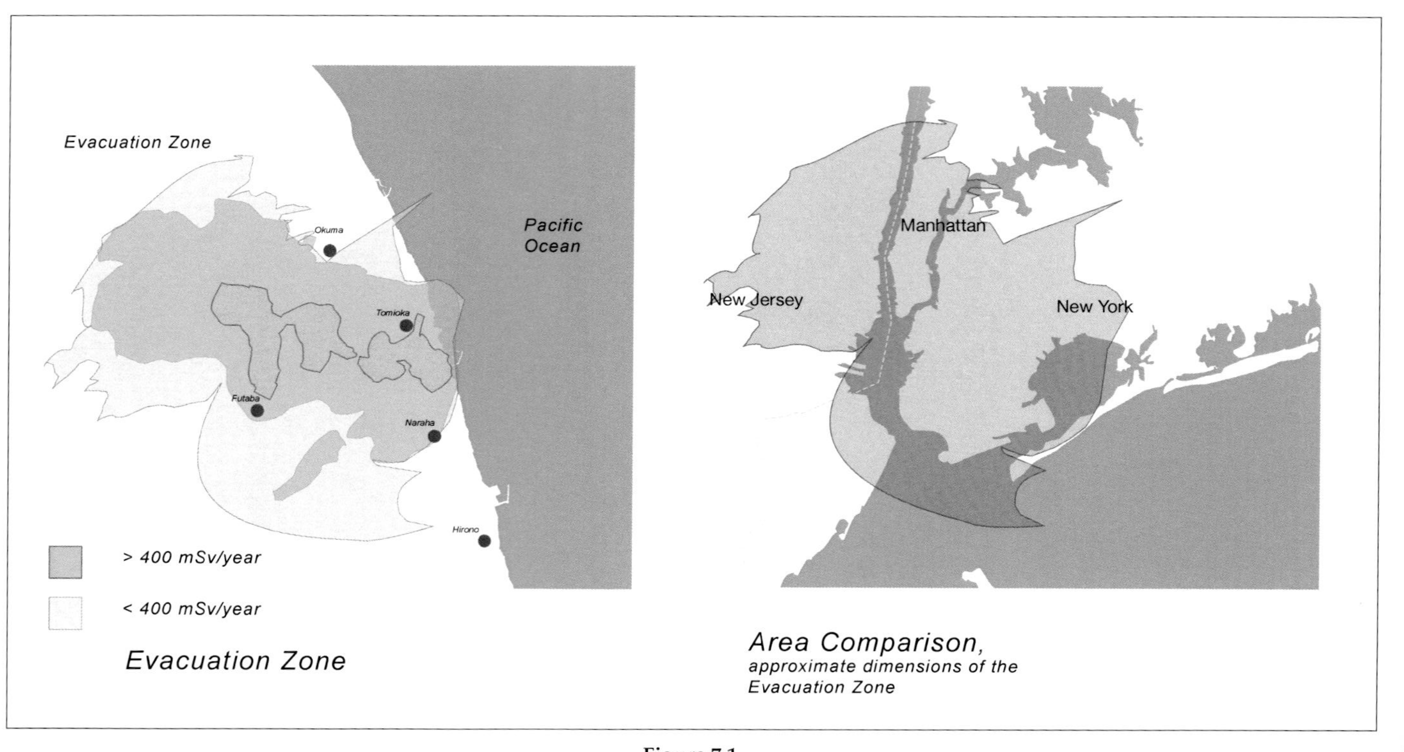
Evacuation Zone
Okuma
Pacific Ocean
Tomioka
Futaba
Naraha
Hirono
> 400 mSv/year
< 400 mSv/year
Evacuation Zone
Manhattan
New Jersey
New York
Area Comparison,
approximate dimensions of the Evacuation Zone

**Figure 7.1**